T0245268

CAMBRIDGE LIBRARY COLLECTION

Books of enduring scholarly value

Mathematical Sciences

From its pre-historic roots in simple counting to the algorithms powering modern desktop computers, from the genius of Archimedes to the genius of Einstein, advances in mathematical understanding and numerical techniques have been directly responsible for creating the modern world as we know it. This series will provide a library of the most influential publications and writers on mathematics in its broadest sense. As such, it will show not only the deep roots from which modern science and technology have grown, but also the astonishing breadth of application of mathematical techniques in the humanities and social sciences, and in everyday life.

Principles of Geometry

Henry Frederick Baker (1866–1956) was a renowned British mathematician specialising in algebraic geometry. He was elected a Fellow of the Royal Society in 1898 and appointed the Lowndean Professor of Astronomy and Geometry in the University of Cambridge in 1914. First published between 1922 and 1925, the six-volume *Principles of Geometry* was a synthesis of Baker's lecture series on geometry and was the first British work on geometry to use axiomatic methods without the use of co-ordinates. The first four volumes describe the projective geometry of space of between two and five dimensions, with the last two volumes reflecting Baker's later research interests in the birational theory of surfaces. The work as a whole provides a detailed insight into the geometry which was developing at the time of publication. This, the third volume, describes the principal configurations of space of three dimensions.

Cambridge University Press has long been a pioneer in the reissuing of out-of-print titles from its own backlist, producing digital reprints of books that are still sought after by scholars and students but could not be reprinted economically using traditional technology. The Cambridge Library Collection extends this activity to a wider range of books which are still of importance to researchers and professionals, either for the source material they contain, or as landmarks in the history of their academic discipline.

Drawing from the world-renowned collections in the Cambridge University Library, and guided by the advice of experts in each subject area, Cambridge University Press is using state-of-the-art scanning machines in its own Printing House to capture the content of each book selected for inclusion. The files are processed to give a consistently clear, crisp image, and the books finished to the high quality standard for which the Press is recognised around the world. The latest print-on-demand technology ensures that the books will remain available indefinitely, and that orders for single or multiple copies can quickly be supplied.

The Cambridge Library Collection will bring back to life books of enduring scholarly value (including out-of-copyright works originally issued by other publishers) across a wide range of disciplines in the humanities and social sciences and in science and technology.

Principles of Geometry

Volume 3: Solid Geometry

H.F. Baker

CAMBRIDGE UNIVERSITY PRESS

Cambridge, New York, Melbourne, Madrid, Cape Town, Singapore,
São Paolo, Delhi, Dubai, Tokyo, Mexico City

Published in the United States of America by Cambridge University Press, New York

www.cambridge.org
Information on this title: www.cambridge.org/9781108017794

This edition first published 1923
This digitally printed version 2010

ISBN 978-1-108-01779-4 Paperback

PRINCIPLES OF GEOMETRY

CAMBRIDGE UNIVERSITY PRESS
C. F. CLAY, MANAGER
LONDON : FETTER LANE, E.C. 4

LONDON : H K. LEWIS AND CO , LTD.,
136, Gower Street, W.C. 1
NEW YORK · THE MACMILLAN CO.
BOMBAY
CALCUTTA } MACMILLAN AND CO , LTD.
MADRAS
TORONTO : THE MACMILLAN CO OF CANADA, LTD.
TOKYO : MARUZEN-KABUSHIKI-KAISHA

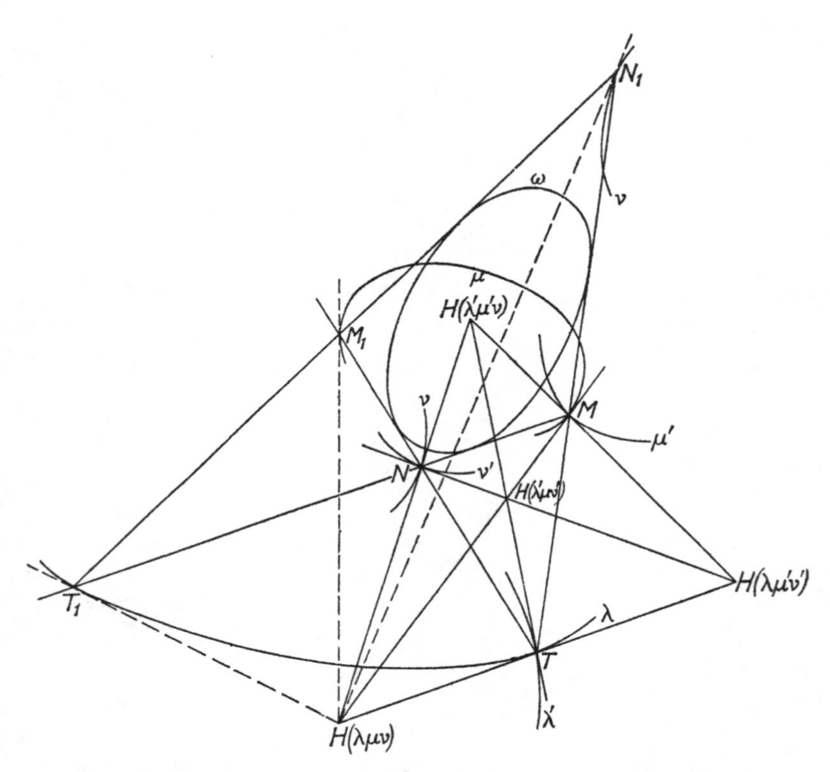

PONCELET'S PORISM AND CONFOCAL QUADRICS
(see p. 116)

PRINCIPLES OF GEOMETRY

BY

H. F. BAKER, Sc.D., LL.D., F.R.S.,

LOWNDEAN PROFESSOR OF ASTRONOMY AND GEOMETRY, AND FELLOW OF
ST JOHN'S COLLEGE, IN THE UNIVERSITY OF CAMBRIDGE

VOLUME III

SOLID GEOMETRY

QUADRICS, CUBIC CURVES IN SPACE, CUBIC SURFACES

CAMBRIDGE
AT THE UNIVERSITY PRESS
1923

PRINTED IN GREAT BRITAIN

PREFACE

THE present volume is devoted to geometry in three dimensions. The discussion of the logical standpoint, to which sufficient space has been given in the preceding volume, is left aside; and, from a desire to limit the size of the volume, many things are omitted which might well have been included. What is given may, however, be regarded as essential to any student who professes to have received a mathematical education. The aptitude for geometrical construction in space, important as it is in the applications of mathematics to physics and engineering, receives, in our educational system at present, less training than it deserves. It is the writer's hope that this volume may help to emphasize this; and may convey to readers something of the fascination and freedom which belongs to the reduction of intricate geometrical relations to the properties of a constructed figure. Only by such methods, moreover, can progress be made beyond the first principles of the subject.

Up to the end of Chapter III, this volume was in type when death severed an association to which the writer owed more help than he can well express. In business, James Bennet Peace was clear and honest; in friendship, constant and self-regardless; many beside the writer deplore his loss. To him, and to the co-operation of the other members of the Staff of the University Press, great acknowledgment is due.

H. F. BAKER.

14 *July* 1923.

"**Teodoro Reye**, *......, che avevo cominciato ad ammirare fin da studente, leggendo la sua classica* Geometrie der Lage; *e col quale poi non avevo tardato ad entrare in relazione scientifica, ed anche personale, sì da poter apprezzare, oltre al valore del matematico, la grande bontà d'animo dell'uomo: vero gentiluomo!*

Nato a Cuxhaven il 20 giugno 1838, ..., era passato verso il 1864 ad insegnare nel Politecnico di Zurigo. ......Aveva esordito nella scienza con lavori di Fisica matematica e di Meteorologia. Ma, poichè a Zurigo il corso del Culmann, fondatore della Statica grafica, si basava sulle teorie della Geometria di posizione, e il classico trattato di Staudt era troppo difficile per gli studenti; Reye fu condotto ad insegnare quelle teorie e ad esporle in un nuovo trattato, che uscì in due parti nel 1866 e nel 1868. ....

....

Artista non meno che scienziato, Reye ha molto contribuito a quella grandiosa e pure snella costruzione scientifica che è la Geometria di posizione, introducendo o svolgendo idee semplici e geniali; studiando, com'è carattere di essa, svariate figure in tal maniera da illuminarne di vivida luce le proprietà più profonde, e i legami che le uniscono. Non solo ci ha fatto conoscere nuovi veri; ma ci ha procurato squisiti godimenti estetici, quali solo può dare il bello. Onore e gratitudine a Lui!"

Corrado Segre, *Rendiconti...dei Lincei*, 2 Aprile 1922.

TABLE OF CONTENTS

CHAPTER I. INTRODUCTION TO THE THEORY OF QUADRIC SURFACES

CHAPTER II. RELATIONS WITH A FIXED CONIC. SPHERES, CONFOCAL SURFACES ; QUADRICS THROUGH THE INTERSECTION OF TWO GENERAL QUADRICS

CHAPTER IV. THE GENERAL CUBIC SURFACE; INTRODUCTORY THEOREMS

CHAPTER I

INTRODUCTION TO THE THEORY OF QUADRIC SURFACES

Preliminary remark. We have, in previous volumes, given trouble to emphasizing the view that the use of the algebraic symbols is not necessary to the geometrical theory; and that the use of symbols of any particular system is equivalent to the adoption of definite geometrical restrictions.

It is, however, often conducive to clearness and brevity, to employ symbols; and it is usual to suppose that the symbols have the same laws of operation as the numbers of ordinary Analysis. Accordingly in the present volume we shall employ such symbols, whenever it seems desirable. The arithmetic notion of the *magnitude* of the symbols, and especially of *infinite* values, remains excluded—as indeed it is in a logical theory of Analysis; and, as heretofore, the *length* and *congruence* of geometrical lines are not employed, save in the conventional sense explained in Volume II (Chap. V).

It will be seen that the utility of the symbols arises chiefly from the use of the *equation of a quadric* (or other) *surface*. We here deduce this from the geometrical definition of the surface. It becomes therefore an interesting problem to replace proofs depending on this equation by direct geometrical deductions from the definition.

Not much space is given in this Volume to the proof, which has already been given in Volume II, p. 191, that a plane is represented by a single equation which is linear in the coordinates; that this is so will be readily understood from the explanation given of the equation of a line, in a plane, in Volume II. A line, in this chapter, is generally given as the join of two points, or the intersection of two planes. The theory of the coordinates of a line is discussed at length in a subsequent section (pp. 56 ff.).

Definition of a quadric surface by means of its lines. We may define a quadric surface, or, as we shall often briefly say, a quadric, in several ways.

Take three arbitrary lines, in space of three dimensions, of which no two intersect. From any point of one of these can be drawn a transversal to meet both the other lines. Consider the aggregate of all the points on all the transversals so drawn. This does not include all points of space, since it is not in general possible to draw from an arbitrary point a transversal to meet three given lines.

The points of the aggregate which lie in an arbitrary plane lie on a conic. For denote the given lines by a, b, c; let an arbitrary plane meet the lines a, b, respectively, in A and B, and let the transversal drawn from an arbitrary point, X, of the third line, c, meet the lines a, b, respectively, in Q and R, and meet the arbitrary plane in P. Then the two axial pencils of planes, having, respectively, a and b as axes, given by the planes $a(X)$ and $b(X)$ as X varies on c, are related. On the arbitrary plane these give therefore the two related pencils of lines, $A(P)$ and $B(P)$. The locus of P is therefore a conic, passing through A and B; this conic equally passes through the point, say C, where the line c meets the arbitrary plane, for evidently a transversal of a and b can be drawn from this point.

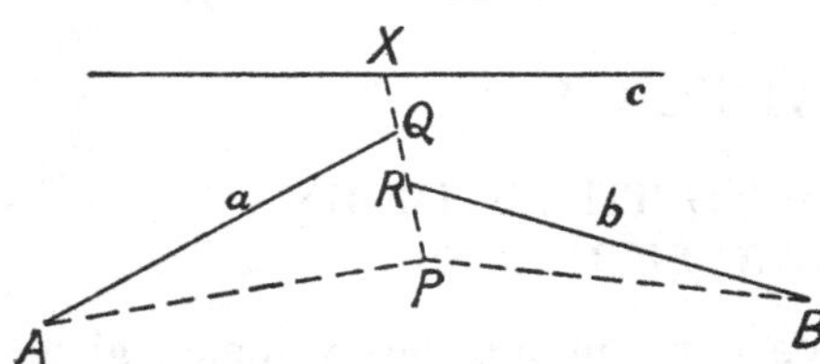

If the points A, B, C are in line, the ray AC of the pencil $A(P)$ is AB, and coincides with the ray BC of the pencil $B(P)$. Then the locus of P consists, beside the line ABC, of a straight line. In other words, the aggregate now being considered, of points lying on all transversals of the three lines a, b, c, contains, in a plane which contains a transversal ABC of a, b, c, beside the points of this line, also the points of another line, say d, which does not meet a or b or c. Every point of this new line, d, is thus a point from which a transversal can be drawn to meet the given lines a, b and c. It is thus clear that the aggregate of points under consideration consists of the points of a system of lines a, b, c, d, ..., infinite in number, of which no two intersect, together with the points of another system of lines, every one of which meets all those of the first system, but of which no two intersect.

This aggregate of points constitutes a quadric. We see that through every point of the locus there pass two lines, consisting of points belonging to the locus, and the plane of these two lines contains no other points of the locus. But a general plane meets the quadric in the points of a conic. Through points $E, A, B, C, D, \ldots$ of this conic will pass lines e, a, b, c, d, ..., no two of which meet, of which the points are points of the quadric, and also lines e', a', b', c', d', ..., also skew to one another, of which every one meets all those of the first system. On the conic, the flat pencil of lines, $E(A, B, C, D)$, is a section of the axial pencil of planes ea', eb', ec', ed', which is met, by any particular one of the lines a, b, c, d, in the points where this line meets the lines a', b', c', d'. Thus we see that these lines a', b', c', d' meet the lines of the other system in related ranges, which are also related to the range, on the conic in a plane section, determined by the points where these lines a', b', c', d' meet the conic. Similarly the lines a, b, c, d determine, on any of the lines e', a', b', c', d', a range related to that of A, B, C, D on the conic.

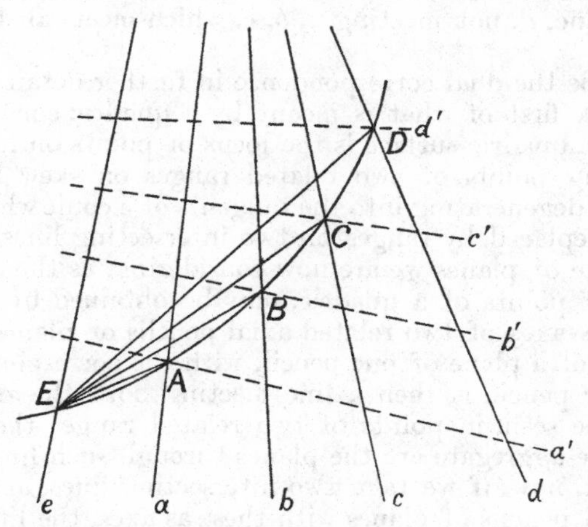

Conversely, if, on two lines, a, b, which do not intersect one another, we have two related ranges, the joins of corresponding points of these ranges are lines which are all met by an infinite number of other lines; the points of these lines are, therefore, those of a quadric surface. For, if P, Q, R be points of the line a, to which correspond the points P', Q', R', respectively, of the line b, and a transversal, c, be drawn from any point of the line PP' to meet QQ' and RR', it follows at once, from what we have seen, that a line drawn from any point of c, to meet a and b, meets these in corresponding points of the two given related ranges thereon. The construction assumes, by hypothesis, that the lines, a, b, of the two related ranges, are not in one plane; if they are, the joins of corresponding points of the two ranges are, as we have seen in Volume II, the tangents of a conic lying in that plane, having the two given lines also as two tangents. The points of a plane, regarded as lying on the tangents of a conic in that plane, may thus be regarded as the points of a degenerate quadric, each tangent line of the conic being taken twice over as a line of the quadric, as a line of each of the systems lying on the quadric.

This account has been obtained by considering the aggregate of the points lying on all the lines which meet three given skew lines. But, in three dimensions, a line is self-dual, and two lines which have a point in common lie also in a common plane; it is proper to consider, then, also the aggregate of all the planes which pass through all the transversals of three given lines. As we have found that through any point of the quadric there pass two lines lying entirely thereon, so we have incidentally found that in any plane passing through a transversal of three given skew lines, a, b, c, there is another line, d, not meeting a, b, c, which meets all transversals of these.

To describe the dual correspondence in further detail it is necessary to speak first of what is meant by a quadric cone. We have seen that the quadric surface is the locus of points on lines joining corresponding points of two related ranges on skew lines, these joining lines degenerating into the tangents of a conic when the skew ranges are replaced by ranges on two intersecting lines. Similarly the aggregate of planes we are now considering, as the dual of the aggregate of points of a quadric, may be obtained by taking two skew lines as axes of two related axial pencils of planes; a line of intersection of a plane of one pencil, with the corresponding plane of the other pencil, is then a line meeting both the axes, and, as may easily be seen, in points of two related ranges thereon. The planes of the aggregate are the planes through such lines of intersection. But, now, if we take two intersecting lines, and have two related axial pencils of planes with these as axes, the line of intersection of a plane of one pencil with the corresponding plane of the other pencil, is a line through the point of intersection of the axes; the aggregate of these lines, passing through a point, constitutes what is called a quadric cone, and is the dual of the aggregate of lines in a plane which touch a conic. As two tangents of the conic meet in a point, so two of the lines forming the cone lie in a plane; as the point of intersection of two tangents of the conic becomes a point of the conic when the two tangents coincide (Vol. II, p. 25), so, if the two lines of the cone coincide, the plane containing them is replaced by a definite plane, called a tangent plane of the cone. The section of a cone by an arbitrary plane is evidently a conic, of which one point is determined by a line of the cone, and a tangent line by a tangent plane of the cone; conversely from any conic we obtain a quadric cone by projection from an arbitrary point not lying in the plane of the conic.

This being understood, consider the dual of the statement that a plane section of a quadric is a conic, through every point of which there passes a line of each of the two systems lying on the quadric. Let an arbitrary point of space be joined to every line of one system of lines lying on the quadric, by a plane. Each of these planes will, as we have seen, contain another line lying on the quadric, of the other system. The aggregate of these planes is then that of the tangent planes of a quadric cone. This is obvious from the fundamental duality of the figure. But it is clear, too, by considering that a variable line, a', of one system, of the quadric surface, meets two lines, a, b, of the other system, in related ranges of points. The plane Oa', joining the line a' to a fixed point O, thus meets the planes Oa, Ob in two related flat pencils of lines, with O as common centre. By taking a section by an arbitrary plane, the planes Oa' give rise to lines in this plane meeting two fixed lines, which are the sections of the arbitrary plane by the planes Oa and Ob, in two related ranges. These lines, therefore, are tangents of a conic in that plane; and the planes Oa' are tangent planes of a quadric cone of which all the lines pass through O. This point O is called the *vertex* of the cone, and its lines are called its *generators*; the lines of a quadric surface are also called the *generators* of the quadric.

A particular consequence is that, as there are two points of the quadric locus lying on an arbitrary line, so there are two of the aggregate of planes, passing through an arbitrary line. These may be obtained, directly from the definition, as the two common corresponding planes of two related axial pencils, just as in the case of the locus of points.

Ex. 1. If A, B, C, D, O be five general points in space, the pairs of planes joining O to the opposite pairs of joins of A, B, C, D, such as OAD, OBC, meet an arbitrary line in three pairs of points which are in involution.

Ex. 2. If an axial pencil of planes be drawn through a line, l, and a related axial pencil of planes be drawn through another line, m, which does not meet l, the lines of intersection, of a plane of the first pencil with the corresponding plane of the second pencil, define a quadric surface, upon which the lines l and m also lie.

Ex. 3. A line l meets the planes BCD, CAD, ABD, ABC, which contain the triads of four arbitrary points A, B, C, D, respectively in P, Q, R, S; another line, l', meets these planes, respectively, in P', Q', R', S'; and the four lines PP', QQ', RR', SS' all lie on the same quadric surface. Prove that the four transversals drawn from A, B, C, D, each to meet both the lines l and l', all lie on a quadric surface. (Cf. Vol. I, p. 30.)

Ex. 4. Let D, P, Q, R be four points in line, and A, B, C be three points whose plane does not contain the line. Let the transversal, l, be drawn from A to meet the lines BQ, CR; the transversal, m, be drawn from B to meet the lines CR, AP; and the transversal, n, be drawn from C to meet the lines AP, BQ. Prove that a line can be drawn from D to meet all of l, m, n.

The representation of this definition by means of the algebraic symbols. We may represent the matter very simply by means of the algebraic symbols. Let a, b, c be three skew lines, of which DAA', BCL, $B'MR$ are any three transversals. Taking D, A, C, B, which do not lie in a plane, as fundamental points, we can choose the symbols of the points A, A', relatively to that of D, so that $A' = D + A$, and, similarly, the symbols of C, L, relatively to that of B, so that $L = B + C$. Then the symbol of R may be supposed to be

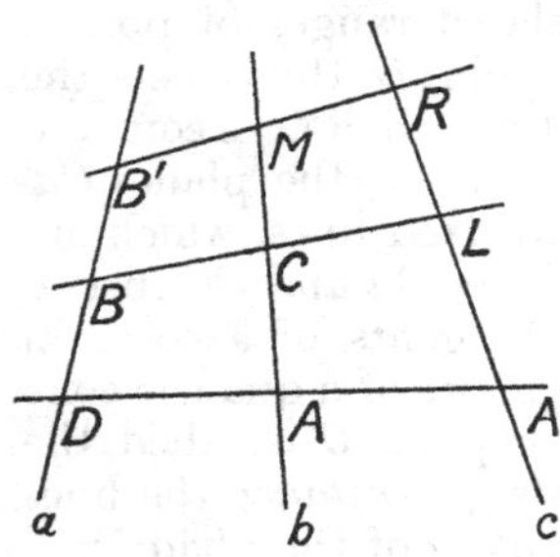

$$R = A' + \mu L = D + A + \mu (B + C)$$
$$= D + \mu B + A + \mu C,$$

where $D + \mu B$ is the symbol of some point on the line a, and $A + \mu C$ of some point on the line b. Hence we infer, for the symbols of B' and M, which lie on a line through R, respectively,

$$B' = D + \mu B, \quad M = A + \mu C.$$

Any other point of the transversal $B'M$ has thus a symbol $B' + \lambda M$, namely

$$\lambda A + \mu B + \lambda\mu C + D,$$

which, then, for different symbols λ, μ, is the general point of the general transversal of the fundamental lines a, b, c. It is thus the general point of the quadric surface under consideration. The symbol can also be written

$$D + \lambda A + \mu (B + \lambda C),$$

and the point lies on a transversal of the lines DA, BC, this transversal meeting these, respectively, in the points $D + \lambda A$, $B + \lambda C$.

We see that every point of the quadric is characterised by two particular algebraic symbols, λ and μ; and further that all points of the quadric for which λ is the same are on a line, not intersecting a, or b, or c; and all points for which μ is the same are on a line, meeting a, b, c but not meeting DA, BC, $B'M$.

The two points of the quadric which lie on an arbitrary line, say on the line joining the points whose symbols are

$$a_1 A + b_1 B + c_1 C + D, \quad a_2 A + b_2 B + c_2 C + D,$$

would then be found by choosing σ, so that the point of this line given by

$$(a_1 + \sigma a_2) A + (b_1 + \sigma b_2) B + (c_1 + \sigma c_2) C + (1 + \sigma) D$$

is the same, for proper values of λ and μ, as the point

$$\lambda A + \mu B + \lambda\mu C + D;$$

this requires

$$(1 + \sigma)(c_1 + \sigma c_2) = (a_1 + \sigma a_2)(b_1 + \sigma b_2).$$

We can express the character of the quadric also by representing any point of it by a symbol

$$XA + YB + ZC + TD,$$

so that X, Y, Z, T are the *coordinates* of this point relatively to the points A, B, C, D; then the sole condition for X, Y, Z, T, is that

$$XY = ZT;$$

this is called the *equation of the quadric*. If we take new points of reference A_1, B_1, C_1, D_1, such that

$$A_1 = C - D, \quad B_1 = A + B, \quad C_1 = A - B, \quad D_1 = C + D,$$

any point of space given by

$$xA_1 + yB_1 + zC_1 + tD_1,$$

is
$$x(C - D) + y(A + B) + z(A - B) + t(C + D),$$

and is a point of the quadric if

$$X = y + z, \quad Y = y - z, \quad Z = t + x, \quad T = t - x,$$

that is, if
$$x^2 + y^2 - z^2 = t^2.$$

This, then, is another form for the equation of the quadric. Whatever θ may be, this equation is satisfied by

$$y + z = \theta^{-1}(t + x), \quad y - z = \theta(t - x);$$

also, whatever ϕ may be, it is satisfied by

$$y + z = -\phi^{-1}(t - x), \quad y - z = -\phi(t + x).$$

It is, however, easy to see that a single *linear* equation connecting the coordinates, (x, y, z, t), of a point, implies that this point lies on a certain plane; for instance, an equation $t = lx + my + nz$ shews that the point, $xA_1 + yB_1 + zC_1 + (lx + my + nz)D_1$ is on the plane through the three points $A_1 + lD_1$, $B_1 + mD_1$, $C_1 + nD_1$. Thus the points for which $y + z = \theta^{-1}(t + x)$, $y - z = \theta(t - x)$ are those of a line; and, by taking different values of θ, we thus obtain a system of lines lying on the quadric surface; it is easy to see that no two of these have a point in common. Another system of mutually non-intersecting lines is given by the equations $y + z = -\phi^{-1}(t - x)$, $y - z = -\phi(t + x)$, for different values of ϕ. It is easily verified that, whatever θ and ϕ may be, the θ-line, of the first system, has a point in common with the ϕ-line, of the second system, this being, in fact, that given by

$$x = \theta + \phi, \quad y = 1 - \theta\phi, \quad z = 1 + \theta\phi, \quad t = \theta - \phi.$$

These expressions then satisfy identically the equation

$$x^2 + y^2 - z^2 - t^2 = 0.$$

Quadric surface defined by two conics in space having two points in common, and a line which meets these conics. We have seen that the points of a quadric which are on a plane lie on a conic. If two points, A, B, of the quadric be taken, and two planes be drawn through these two points, we thus obtain two conics lying on the quadric. Considering only the lines of the quadric belonging to one system of generators, there will be one line of this system through an arbitrary point, P, of one of these conics; this line will meet the plane of the other conic in a point, say P', which, as the line lies entirely on the quadric, will be on the other conic. There is thus a (1, 1) correspondence between the points of the two conics, whereby to every point, P, of one of these conics, there corresponds a definite point, P', of the other, while P' equally determines P. By what has been proved above, the range of points, P', on one conic, is related to the range of points, P, on the other.

Conversely, let two arbitrary conics be given, lying in different planes, but having two points, A, B, in common; let a further arbitrary fixed point, C, be taken on one conic, and a point, C', on the other; then, we can, to any variable point, P, of the former conic, make correspond a definite point, P', of the latter conic, by the condition that the range of points, A, B, C, P, of the former conic is related to the range of points, A, B, C', P', of the latter. The aggregate of all the points lying on all the lines PP' is then a quadric surface. And when the conics only are given, there is an infinite number of quadrics so obtainable, since we can choose the point C', which is to correspond to C, in an infinite number of ways. A definite quadric is determined by the two conics together with the line CC'. In particular, if the tangents of one conic at the points A, B meet in the point T, and the tangents of the other conic at the points A, B meet in the point U, and the line CC' intersect the line TU, say in O, it can be shewn that the conics are in perspective from the point O. For if the lines which join the point O to the points, P', of one of the two given conics, σ', be allowed to meet the plane of the other conic, σ, in points, Q, it follows, because the pencils of lines, $A(P')$, $B(P')$, are related, that the pencils of lines, $A(Q)$, $B(Q)$, are also related, so that Q describes a conic, τ, in the plane of the conic σ. As O, A, T, U are in one plane, the conic τ will touch the conic σ at A; similarly it will touch it at B; and the point C, lying on OC', will be common to the conics τ and σ. Two conics which touch at two points, and have another point in common, are however coincident. Thus the lines, PP', joining corresponding points of the two conics, all pass through O. The quadric determined by this particular correspondence of C' to C thus reduces to a quadric cone, of vertex O.

In symbols, referred to A, B, T, U, the corresponding points P, P' of the two conics may be taken to be

$$P = \theta^2 A + \theta T + B, \quad P' = \phi^2 A + \phi U + B,$$

where θ, ϕ, by their variation, are to give two related ranges on the two conics, with the condition that the values $\theta^{-1} = 0$, $\theta = 0$, at

A, B, are to correspond to the values $\phi^{-1}=0$, $\phi=0$; thus ϕ is of the form θa^{-1}, where a is the same for all points P, P', but depends upon the initial correspondence adopted, of C' to C. A general point $\rho P + aP'$, of the line PP', is then

$$\theta^2(\rho+a^{-1})A+\theta\rho T+\theta U+(\rho+a)B,$$

or, say, $xA+tT+zU+yB$, so that

$$x=\theta^2(\rho+a^{-1}),\quad t=\theta\rho,\quad z=\theta,\quad y=\rho+a;$$

eliminating ρ we have, for the equation satisfied by the coordinates of any point of the line PP',

$$xy=z^2+(a+a^{-1})zt+t^2;$$

if we put

$$\xi=x+y,\quad \eta=x-y,\quad \zeta=(a^{\frac{1}{2}}+a^{-\frac{1}{2}})(z+t),\quad \tau=(a^{\frac{1}{2}}-a^{-\frac{1}{2}})(z-t),$$

this equation is the same as

$$\xi^2-\eta^2-\zeta^2+\tau^2=0,$$

which is another form of the equation of the quadric.

But, without assigning the correspondence of the two points C, C', we may suppose the two conics to be given, respectively, with the coordinates here used, by the equations

$$z=0,\ xy-t^2=0,\quad \text{and}\quad t=0,\ xy-z^2=0;$$

then, whatever k may be, the equation

$$xy=z^2+2kzt+t^2,$$

is evidently that of a locus containing both the given conics. And k can be chosen so that the locus contains an arbitrary point of space which does not lie in the plane of either of the two conics. A plane drawn through such an arbitrary point, say O, will meet each of the two conics in two points; through the four points so found, and the point O, a definite conic can be drawn. The points of all such conics, obtained by drawing different planes through the same point O, do in fact constitute a quadric determined by the two conics and the point O. To prove this, it is sufficient, after what has preceded, to shew that a quadric can be found to contain two conics which have two points in common, and to contain also an arbitrary point not lying on the plane of either of the two conics; and this follows from the equation just put down. But we can deduce the result at once without the symbols, from the preceding theory, by remarking that a line (two lines, indeed) can be drawn from the arbitrary point to meet the given conics, say in C and C' respectively, and these points C, C' can then be used to establish a (1, 1) relation of points of the two conics. That such a transversal of the two conics can be drawn from the arbitrary point, is clear by projecting one of the conics, from the point, on to the plane of the other conic; thereby a conic is obtained having, beside A and B, two points of intersection with the given conic on that plane. Thereby two such transversals are obtained; but these lead to the same quadric, being the generators of the two systems of the quadric which pass through the arbitrary point.

When k is real, in the quadric given by the equation

$$xy = z^2 + 2kzt + t^2,$$

there is a difference according to the value of k, that is, according to the position of the arbitrary point, which is given in addition to the two conics. When $k^2 > 1$, there exists a *real* number a such that $2k = a + a^{-1}$; then the quadric contains real lines, as we have seen. When $k^2 < 1$, the lines are imaginary. When $k^2 = 1$, the equation of the quadric surface is one of the two represented by

$$xy = (z + t)^2, \quad xy = (z - t)^2,$$

and the surface is, in fact, a cone. In this case the corresponding points, P, P', of the two conics, are

$$P = \theta^2 A + \theta T + B, \quad P' = \theta^2 A \pm \theta U + B,$$

and the line PP' contains one of the two points $T \mp U$, whatever P and P' may be.

An interesting simple result follows from what we have said; *if we have three conics in space of which every two have two points in common, then a quadric surface exists containing all these conics.* This remark is made by Poncelet, *Traité des propriétés projectives des figures*, 1865, I, p. 378, § 606.

Ex. 1. There are in fact just two quadric cones which can be constructed to contain two conics in space which have two points

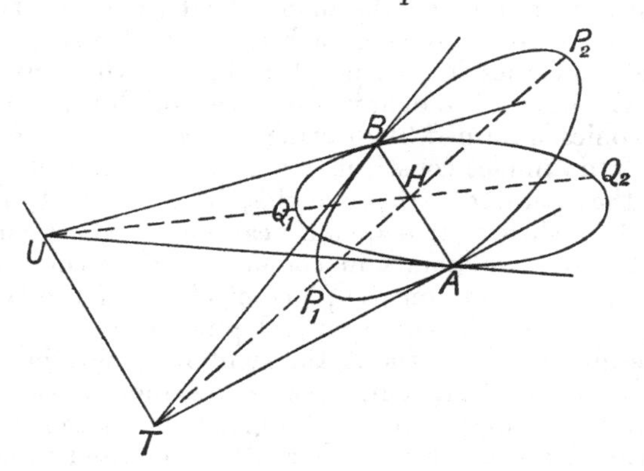

in common, as will appear also in another way. We may construct these, if T, U be, as above, the intersections of the tangents of the two conics, respectively, at their common points A, B, by taking an arbitrary point, say H, upon the line AB, then joining TH, to meet the conic of the plane TAB in P_1 and P_2, and joining UH, to meet the conic of the plane UAB in Q_1 and Q_2. It is then clear that the lines P_1Q_1, P_2Q_2 meet in a point, say O, of the line TU; and the lines P_1Q_2, P_2Q_1 meet in another point, say O', of this line. The lines joining O to the points of one of the conics meet the plane of the other conic in this conic itself, and constitute one of the two quadric cones, with O as vertex; and O' is the vertex of the other cone. The same points, O and O', are found whatever be the point H taken on AB.

Ex. 2. Shew that the equations of three conics in space of which every two have two points in common, can be taken to be, respectively,

$$x=0,\ y^2+2fyz+z^2=t^2;\quad y=0,\ z^2+2gzx+x^2=t^2;$$
$$z=0,\ x^2+2hxy+y^2=t^2.$$

All these then lie on the locus represented by the equation

$$x^2+y^2+z^2+2fyz+2gzx+2hxy=t^2.$$

Definition of a quadric by two related central systems of lines and planes. We have in Vol. I (pp. 148, 149) explained what is meant by two related plane systems of points and lines, of which the correspondence is determined by assigning four points of one plane to correspond respectively to four points of the other, no three of the four points being in line. In this relation, to any point of one plane corresponds a point of the other, the points of a range, on a line of one plane, corresponding to the points of a related range on a corresponding line of the other plane; and, consequently, to a pencil of lines through a point in one plane, corresponds a related pencil of lines through the corresponding point of the other plane. We may, however, have two dually related plane systems, in which to a point of one plane corresponds a line of the other plane, and, to the points of a range of the former plane, correspond the lines of a pencil of lines through the corresponding point of the latter plane; then to the lines of a pencil in the former plane will correspond the points of a range in the latter plane. As we have established the theory of related ranges and pencils, in Vol. I, by means of incidences, coupled with Pappus' theorem, which is a self-dual theorem, it follows at once that when a range in the former plane corresponds to a pencil in the latter plane, these will be related to one another; and, hence, that when a pencil of lines in the former plane corresponds to a range of points in the latter plane, these will also be related.

When we have, in space of three dimensions, two related plane systems, ϖ and ϖ', and, also, two points, O and O', of which O does not lie in the plane ϖ, and O' does not lie in the plane ϖ', we may join the points and lines of ϖ to O, by lines and planes, respectively, and the points and lines of ϖ' to O', by lines and planes, respectively. Thereby we obtain two systems of lines and planes, passing respectively through O and O', which we may call *star systems*, or *central systems*. When the plane systems ϖ and ϖ' are related, we may speak of these central systems as being related.

But then, equally, if we have two dually related plane systems, ϖ and σ', and two points, O and O', of which O is not in the plane of ϖ, and O' is not in the plane of σ', we may join the points and lines of ϖ to O by lines and planes, respectively, and the lines and points of σ' to O' by planes and lines, respectively. Then we obtain two central systems which are dually related. To a line of the system of centre O, will correspond a plane of the system of centre O': to a plane through O, containing two lines of the system O, will correspond the line through O', which is the intersection of the planes through O' corresponding to the lines through O; to a pencil of lines through O, lying in a plane through O, will correspond an axial pencil of planes through O', all passing through the line, through O', which corresponds to the plane of the flat pencil through O; and this axial pencil of planes will be related to the flat pencil of lines. And the same may be said of the correspondence, of planes and lines through O', to the lines and planes through O.

Taking now two such dually related central systems, of centres O and O', we consider the locus of the intersection of any line of the system O with the corresponding plane of the system O'. It is the fact that this locus is a quadric surface containing O and O'; conversely, on a given quadric surface we may take two points O and O', arbitrarily, and, in an infinite number of ways, regard the surface as the locus of the intersection of lines of a central system of centre O, with the corresponding planes of a dually related system of centre O'. The theory arises naturally in a later section, devoted to the (1, 1) correspondence of two spaces, and we shall limit ourselves here to the proof of the first part of the statement we have made. To every line, l, through O, there corresponds a plane, λ', through O'; and to every line in the plane λ', passing through O', there corresponds a plane λ, passing through the line l, and through O; and the same for O' and O as for O and O'. In particular, to the line $O'O$, of the system O', there corresponds a plane τ, through O; to every line, l, through O, in this plane τ, there corresponds then a plane through the line $O'O$. This plane, which corresponds to the line l, will meet the plane τ in a line, say m, passing through O. We shall thus have two related pencils of lines, l and m, in the plane τ, all passing through O. These two related pencils will, in the case we regard as general, have two common corresponding rays.

There are, therefore, two lines l, through O, in the plane τ, such that the corresponding planes of the central system O', respectively, contain them; we may call these lines g_1 and g_2, the planes through O' to which they correspond being called, respectively, γ_1' and γ_2'. As every point of the line g_1 lies in the corresponding plane γ_1', every point of g_1 is a point of the locus; and, similarly, so is every point of g_2. Then, to every plane through g_1 (and hence, through O), will correspond a line through O' lying in the plane γ_1'; in particular, to the plane g_1O', will correspond a line g_2', through O', which, as it lies in this plane g_1O', will meet g_1. And there will, similarly, be a line g_1', through O', in the plane g_2O', and, therefore, meeting g_2. The lines g_2', g_1' will be the lines through O' which lie, respectively, in the planes through O which correspond to them.

Now consider a line meeting g_1, say in N, and meeting g_1', say in N'. Then, to the lines through O in the plane NON', correspond planes through O', all passing through a line, and forming an axial pencil related to the pencil of lines through O. The lines in question, and these planes, will meet the line NN' in two related ranges; in the most general case these ranges will have two corresponding points in common; these are the points N, N', and no other point of the locus lies on the line NN'. The locus, however, contains a point on any line drawn through O; let P be such a point, not lying on g_1 or g_1'; let PNN' be the transversal from P to g_1 and g_1'. Then, upon this line, there are two related ranges, as explained, which now have three common corresponding points, and therefore coincide entirely. The line PNN' thus lies entirely on the locus. By taking various points P we can thus obtain an infinite number of lines such as PNN', lying entirely on the locus. If three such lines be taken, it can be proved that any transversal of these lies entirely on the locus, each of these transversals containing three corresponding points of two related ranges in common.

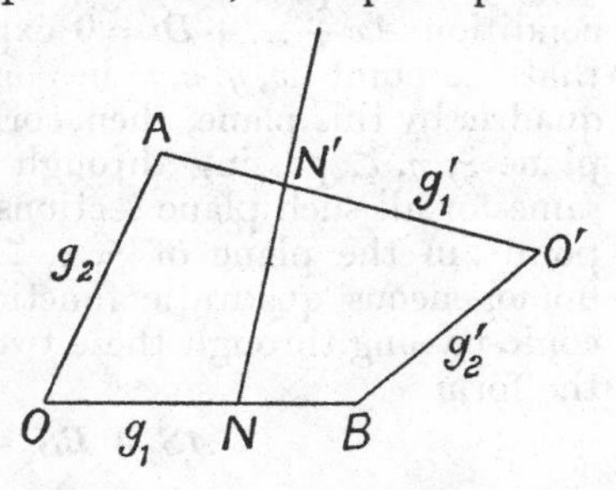

The locus is thus identified with that of all points on all the transversals of three skew lines, namely as a quadric.

Quadric as representing a plane on which two fundamental points are given. The expression above given of the points of a quadric by means of two parameters

$$x = \theta + \phi, \quad y = 1 - \theta\phi, \quad z = 1 + \theta\phi, \quad t = \theta - \phi,$$

is capable of being regarded from a point of view to which brief reference may now be made. We have already, in Vol. II (p. 191), come to regard the points of a plane on which two absolute points are given as arising from the points of a quadric. Without repeating what is there said, we refer here to a general point of view.

If for the parameters θ, ϕ, in the formulae just cited, we write, respectively, ξ/ζ and η/ζ, we may suppose the points of the quadric to be given by

$$x = \zeta(\xi + \eta), \quad y = \zeta^2 - \xi\eta, \quad z = \zeta^2 + \xi\eta, \quad t = \zeta(\xi - \eta);$$

if we regard ξ, η, ζ as homogeneous coordinates in a plane, there will correspond, in this plane, to a condition $Ax + By + Cz + Dt = 0$ for the points of the quadric, the equation

$$A\zeta(\xi + \eta) + B(\zeta^2 - \xi\eta) + C(\zeta^2 + \xi\eta) + D\zeta(\xi - \eta) = 0;$$

this is easily seen to be that of the general conic, in the plane of ξ, η, ζ, which passes through the two points $(1, 0, 0)$, $(0, 1, 0)$. The condition $Ax + \ldots + Dt = 0$ expresses, as we have remarked above, that the point (x, y, z, t) lies on a certain plane; the section of the quadric by this plane, then, corresponds to a particular conic in the plane ξ, η, ζ, passing through two definite points, which are the same for all such plane sections. More generally, take two absolute points in the plane of $\xi. \eta, \zeta$, and let S_1, S_2, S_3, S_4 be any four homogeneous quadratic functions of ξ, η, ζ, such that the general conic passing through these two absolute points has an equation of the form

$$AS_1 + BS_2 + CS_3 + DS_4 = 0;$$

then, to any point (ξ, η, ζ) of the plane, determine a point in three-fold space whose coordinates are given by

$$\frac{x}{S_1} = \frac{y}{S_2} = \frac{z}{S_3} = \frac{t}{S_4};$$

the coordinates x, y, z, t will then be connected by a rational homogeneous equation, obtainable by elimination of the two ratios ξ/ζ, η/ζ from the three equations by which the ratios of x, y, z, t are defined. This equation may be regarded as the equation of a locus. Any general line will meet this locus in two points; for such a line consists of points, whose coordinates satisfy the equations of two planes, $Ax + \ldots + Dt = 0$, and $A'x + \ldots + D't = 0$; and these correspond to points (ξ, η, ζ) lying on two conics $AS_1 + \ldots + DS_4 = 0$, $A'S_1 + \ldots + D'S_4 = 0$; as these conics have two points in common, which are independent of $A, \ldots, D, A', \ldots, D'$, they will have two other points in common. It is in fact the case that this locus is a quadric surface, as will appear abundantly below. To any point of this, which lies on an infinite number of planes, will correspond a single point of the plane (ξ, η, ζ).

Consideration of the most general homogeneous equation of the second order connecting the space coordinates. It will add considerably to clearness to prove at once that the most general homogeneous quadratic equation connecting x, y, z, t implies that the point x, y, z, t lies upon a quadric surface, or upon a locus which we may regard as a particular case of this.

Write this equation in the form

$$ax^2 + by^2 + cz^2 + dt^2 + 2fyz + 2gzx + 2hxy + 2uxt + 2vyt + 2wzt = 0,$$

which we may abbreviate into either of the forms

$$(a, b, c, d, f, g, h, u, v, w \between x, y, z, t)^2 = 0, \quad (x, y, z, t)^2 = 0;$$

we prove, first, that, in an infinite number of ways, this can be put into the form

$$A_1(a_1x + b_1y + c_1z + d_1t)^2 + A_2(a_2x + b_2y + c_2z + d_2t)^2$$
$$+ A_3(a_3x + b_3y + c_3z + d_3t)^2 + A_4(a_4x + b_4y + c_4z + d_4t)^2 = 0,$$

wherein $A_1, A_2, A_3, A_4, a_1, b_1, c_1, d_1, \ldots, a_4, b_4, c_4, d_4$ are all rational real functions of the coefficients, $a, b, c, d, \ldots, w$, which occur in the original equation.

1°. A quadratic form in two variables, x and y, say $ax^2+2hxy+by^2$, if the coefficient b be not zero, is

$$b^{-1}(ab - h^2)x^2 + b^{-1}(hx + by)^2;$$

and similarly if a be not zero. If a and b be both zero it is

$$\tfrac{1}{2}h[(x+y)^2 - (x-y)^2].$$

In any case, therefore, it is of the form $A\xi^2 + B\eta^2$, where A, B are rational real functions of a, b, h, and ξ, η are linear functions of x, y with coefficients which are real rational functions of a, b, h.

2°. A quadratic form in three variables, x, y, z, say

$$ax^2 + by^2 + cz^2 + 2fyz + 2gzx + 2hxy,$$

if one of the coefficients a, b, c be not zero, for example, if c be not zero, can be written

$$c^{-1}(gx + fy + cz)^2 + c^{-1}[(ca - g^2)x^2 - 2(fg - ch)xy + (bc - f^2)y^2],$$

wherein the terms after the square constitute a quadratic form in the variables x, y only; if a, b, c be all zero, and none of f, g, h be zero, the form is

$$\tfrac{1}{2}fg^{-1}h^{-1}(gz + hy + 2f^{-1}ghx)^2 - 2f^{-1}ghx^2 - \tfrac{1}{2}fg^{-1}h^{-1}(gz - hy)^2;$$

if a, b, c be all zero, and one of f, g, h be zero, or two of these, for example either $f = 0$, or $f = g = 0$, the form is, respectively, $2x(gz + hy)$ or $2hxy$, that is, respectively,

$$\tfrac{1}{2}[(x + gz + hy)^2 - (x - gz - hy)^2], \quad \tfrac{1}{2}h[(x+y)^2 - (x-y)^2].$$

In any case then the form can be written as $A\xi^2 + B\eta^2 + C\zeta^2$, where A, B, C are real rational functions of the original coefficients $a, b, \ldots, h$, of which one, or two, may be zero, and ξ, η, ζ are independent linear functions of x, y, z, with coefficients which are rational real functions of $a, b, \ldots, h$.

3°. The quadratic form in four variables,

$$ax^2 + by^2 + cz^2 + dt^2 + 2fyz + 2gzx + 2hxy + 2uxt + 2vyt + 2wzt,$$

(i) if not all of a, b, c, d be zero, for example if d be not zero, is

$$d^{-1}(ux + vy + wz + dt)^2 + \phi,$$

where ϕ is a quadratic form in x, y, z only, whose coefficients are rational real functions of the original coefficients $a, b, \ldots, w$.

(ii) if all of a, b, c, d be zero, but none of f, g, h be zero, we can choose l, m, n so that

$$hm + gn + u = 0, \quad hl + fn + v = 0, \quad gl + fm + w = 0,$$

namely

$$l = \tfrac{1}{2}g^{-1}h^{-1}(fu - gv - hw), \quad m = \tfrac{1}{2}h^{-1}f^{-1}(-fu + gv - hw),$$
$$n = \tfrac{1}{2}f^{-1}g^{-1}(-fu - gv + hw);$$

then, substituting in the quadratic form, for x, y, z, respectively,

$$x = x' + lt, \quad y = y' + mt, \quad z = z' + nt,$$

this form reduces to

$$2fy'z' + 2gz'x' + 2hx'y' + 2t^2(mnf + nlg + lmh + ul + vm + wn),$$

wherein the first part is a quadratic form in x', y', z' only, and the last part contains only the term in t^2.

(iii) if all of a, b, c, d be zero, and f be zero, but not both of g and h, the form is

$$2(hy + gz)x + 2(ux + vy + wz)t;$$

here it may happen that $ux + vy + wz$ is of the form $ux + \lambda(hy + gz)$, namely that $v/h = w/g$, $(= \lambda)$; then the form is

$$2(hy + gz)x + 2[ux + \lambda(hy + gz)]t,$$

and is a quadratic form in the *three* variables $x, t, hy + gz$; if $ux + vy + wz$ is not of the form in question, taking ξ, η, ζ, τ so that

$$hy + gz = \eta + \zeta, \quad ux + vy + wz = \tau + \xi,$$
$$x = \eta - \zeta, \qquad t = \tau - \xi,$$

we can, from these, determine x, y, z, t as linear functions of ξ, η, ζ, τ, with coefficients which are real and rational in the coefficients of the original form, which then is

$$2(\eta^2 - \zeta^2) + 2(\tau^2 - \xi^2).$$

(iv) if all of a, b, c, d be zero, and all of f, g, h be zero, the form is

$$2(ux + vy + wz)\,t,$$

and is a quadratic form in the *two* variables $ux + vy + wz$, t.

Thus in all cases the reduction is possible, to the form

$$A_1(a_1x + b_1y + c_1z + d_1t)^2 + \ldots + A_4(a_4x + b_4y + c_4z + d_4t)^2,$$

wherein the four linear forms are linearly independent.

Consideration of particular cases. The case of a cone. We regard the case in which none of the coefficients A_1, A_2, A_3, A_4 is zero as the general case. Before coming to this we consider the particular cases, and, first, the case in which one of these coefficients, and only one, say A_4, is zero. Then if we put

$$\xi = a_1x + b_1y + c_1z + d_1t, \quad \eta = a_2x + b_2y + c_2z + d_2t,$$
$$\zeta = a_3x + b_3y + c_3z + d_3t, \quad \tau = a_4x + b_4y + c_4z + d_4t,$$

which is equivalent to choosing for the points of reference, for the space, the four points of intersection of the four planes which are represented by the equations $\xi = 0, \eta = 0, \zeta = 0$ and $\tau = 0$, the fundamental quadratic form becomes $A_1\xi^2 + A_2\eta^2 + A_3\zeta^2$. We consider then what is represented by the equation

$$A_1\xi^2 + A_2\eta^2 + A_3\zeta^2 = 0.$$

Let $(\xi, \eta, \zeta, 1)$ be any point which satisfies this equation; the general point on the line which joins this to the point $\xi = 0$, $\eta = 0$, $\zeta = 0$, $\tau = 1$, has coordinates of the form $(\xi, \eta, \zeta, 1 + \lambda)$; this point then, also, satisfies the equation. The equation is then satisfied by all the points of an aggregate of lines passing through the point $(0, 0, 0, 1)$. These lines evidently meet the plane $\tau = 0$ in the points of a conic, of which the equations are $\tau = 0$, $A_1\xi^2 + A_2\eta^2 + A_3\zeta^2 = 0$. The equation therefore represents what we have called a quadric cone. Conversely, from the definition of such a cone given above, it follows that its equation must be capable of this form.

Thus we see that the necessary and sufficient condition that the general quadratic polynomial, in the space coordinates x, y, z, t, should represent a quadric cone, when equated to zero, is that this polynomial should be capable of being expressed as a function of only three linear functions of x, y, z, t.

We can hence obtain the condition for this in terms of the coefficients in the original form. For, if this be $f(x, y, z, t)$, and be capable of being expressed in terms of the three linear functions, ξ, η, ζ, of x, y, z, t, in the form $f(x, y, z, t) = F(\xi, \eta, \zeta)$, we have four identities such as

$$\frac{\partial f}{\partial x} = \frac{\partial \xi}{\partial x} \cdot \frac{\partial F}{\partial \xi} + \frac{\partial \eta}{\partial x} \cdot \frac{\partial F}{\partial \eta} + \frac{\partial \zeta}{\partial x} \cdot \frac{\partial F}{\partial \zeta},$$

wherein the factors $\partial\xi/\partial x$, $\partial\eta/\partial x$, etc., are functions of the original coefficients alone, and do not depend on x, y, z, t. Thus the four linear functions of x, y, z, t given by $\partial f/\partial x$, $\partial f/\partial y$, etc., are all expressible as linear functions of the three forms $\partial F/\partial\xi$, $\partial F/\partial\eta$, etc. which are linear in ξ, η, ζ and therefore, also, in x, y, z, t. There exist, however, values of x, y, z, t reducing to zero these three linear forms $\partial F/\partial\xi$, etc. There exist, therefore, values of x, y, z, t reducing to zero the four forms $\partial f/\partial x$, etc. These forms are, respectively,

$$2(ax + hy + gz + ut), \quad 2(hx + by + fz + vt),$$
$$2(gx + fy + cz + wt), \quad 2(ux + vy + wz + dt),$$

and the condition necessary and sufficient that there should be values of x, y, z, t for which these all vanish is that the determinant

$$\Delta, = \begin{vmatrix} a, & h, & g, & u \\ h, & b, & f, & v \\ g, & f, & c, & w \\ u, & v, & w, & d \end{vmatrix},$$

should be zero.

Conversely, if this determinant is zero, and (x_0, y_0, z_0, t_0) be values, then existing, for which all of $\partial f/\partial x$, $\partial f/\partial y$, etc., vanish, while (x, y, z, t) are any values for which $f(x, y, z, t) = 0$, then every point on the line joining the point (x_0, y_0, z_0, t_0) to (x, y, z, t) is a point for which $f(x, y, z, t) = 0$. For such a point has coordinates of the form $(x_0 + \lambda x, y_0 + \lambda y, z_0 + \lambda z, t_0 + \lambda t)$, and we have

$$f(x_0 + \lambda x, \ldots) = f(x_0, y_0, z_0, t_0)$$
$$+ \lambda\left(x\frac{\partial f}{\partial x_0} + y\frac{\partial f}{\partial y_0} + z\frac{\partial f}{\partial z_0} + t\frac{\partial f}{\partial t_0}\right) + \lambda^2 f(x, y, z, t);$$

now, by hypothesis, $f(x, y, z, t) = 0$; and $\partial f/\partial x_0 = 0$, $\partial f/\partial y_0 = 0$, etc., and, therefore, also $f(x_0, y_0, z_0, t_0) = 0$, since

$$f(x_0, y_0, z_0, t_0) = \frac{1}{2}\left(x_0\frac{\partial f}{\partial x_0} + \ldots + t_0\frac{\partial f}{\partial t_0}\right).$$

It follows from this that the equation $f(x, y, z, t) = 0$ is satisfied by all points of an aggregate of lines passing through the point (x_0, y_0, z_0, t_0). It is also easy to see directly that, with the hypothesis made, namely that the four forms $\partial f/\partial x$, $\partial f/\partial y$, etc., are all annulled by the values $x = x_0$, $y = y_0$, etc., the quadratic form $f(x, y, z, t)$ is a function of three forms, linear in x, y, z, t; which, when t_0 is not zero, may be taken to be $xt_0 - x_0t$, $yt_0 - y_0t$, $zt_0 - z_0t$. We find at once, indeed, when we utilise this hypothesis, that

$$a(xt_0 - x_0t)^2 + b(yt_0 - y_0t)^2 + c(zt_0 - z_0t)^2 + 2f(yt_0 - y_0t)(zt_0 - z_0t)$$
$$+ 2g(zt_0 - z_0t)(xt_0 - x_0t) + 2h(xt_0 - x_0t)(yt_0 - y_0t)$$

is equal to $t_0^2 f(x, y, z, t)$. This supposes that t_0 is not zero; but we can suppose that one at least of x_0, y_0, z_0, t_0 is not zero, and so, in any case, obtain a similar formula.

The case of two planes. Consider now the more particular case when the quadratic form in x, y, z, t, $f(x, y, z, t)$, reduces to only two squares,

$$A_1(a_1x + b_1y + c_1z + d_1t)^2 + A_2(a_2x + b_2y + c_2z + d_2t)^2,$$

say to $A_1\xi^2 + A_2\eta^2$. The equation $f(x, y, z, t) = 0$ is then equivalent to

$$(\sqrt{A_1}\,\xi + \sqrt{-A_2}\,\eta)(\sqrt{A_1}\,\xi - \sqrt{-A_2}\,\eta) = 0,$$

so that, as ξ and η are linear in x, y, z, t, it represents two planes. These are imaginary when A_1 and A_2 are real and both positive, or negative. In this case, by identities such as those used in the last case, the four linear forms $\partial f/\partial x$, $\partial f/\partial y$, etc., are linearly expressible by *two* linear forms; thus every first minor, of three rows and columns, in the determinant

$$\Delta, = \begin{vmatrix} a, & h, & g, & u \\ h, & b, & f, & v \\ g, & f, & c, & w \\ u, & v, & w, & d \end{vmatrix},$$

vanishes. Conversely, denote respectively by U, V, W, P the linear forms in x, y, z, t, $\frac{1}{2}\partial f/\partial x$, $\frac{1}{2}\partial f/\partial y$, $\frac{1}{2}\partial f/\partial z$, $\frac{1}{2}\partial f/\partial t$, and assume that the three particular first minors of this determinant which are the cofactors respectively of c, d and w, are all zero (from which, as will appear, all the first minors are zero), and that the second minor $ab - h^2$ is not zero; then it is easy to verify that, for all values of x, y, z, t, the forms W and P are expressible in terms of U and V, by means of the equations

$$-(ab - h^2)W = (hf - bg)U + (gh - af)V,$$
$$-(ab - h^2)P = (hv - bu)U + (hu - av)V;$$

from these it follows that $(ab - h^2)f(x, y, z, t)$, which is equal to

$$(ab - h^2)(xU + yV + zW + tP),$$

is equal to $U[(ab - h^2)x + (bg - hf)z + (bu - hv)t]$
$+ V[(ab - h^2)y + (af - gh)z + (av - hu)t]$

and hence to $U(bU - hV) + V(aV - hU)$,

or to $aV^2 - 2hUV + bU^2$.

Wherefore, $f(x, y, z, t)$ is expressible in the form $A_1\xi^2 + A_2\eta^2$, where ξ, η are linear in x, y, z, t. This same conclusion follows, in a similar way, whenever all the first minors of the determinant Δ are zero, if there be one second minor which is not zero.

The case of two coincident planes. For the last particular case suppose that $f(x, y, z, t)$ reduces to the single term

$$A_1(a_1x + b_1y + c_1z + d_1t)^2.$$

Then, we see, as in the preceding cases, that the four linear forms $\partial f/\partial x$, $\partial f/\partial y$, etc., are identical save for a factor independent of x, y, z, t. When this is so, every second minor of the determinant Δ vanishes. Conversely suppose that this is so, but $f(x, y, z, t)$ does not vanish identically. The forms U, V, W, P, with the notation just employed, are then the same, save for a factor independent of x, y, z, t. The coefficients a, b, c, d, respectively of x^2, y^2, z^2, t^2, in $f(x, y, z, t)$, cannot all vanish, since then, from $bc - f^2 = 0$, $ad - u^2 = 0$, would follow $f = 0$, $u = 0$, and similarly all other coefficients in $f(x, y, z, t)$ would vanish. Suppose then, for example, that a is not zero; then from the equivalence of the forms U, V, W, P, it follows that $U/a = V/h = W/g = P/u$, and hence

$$f(x, y, z, t) = xU + yV + zW + tP = a^{-1}U(ax + hy + gz + ut) = a^{-1}U^2,$$

so that $f(x, y, z, t)$ is the square of a single linear form, and the equation $f(x, y, z, t) = 0$ represents a plane taken twice over.

After Sylvester (*Coll. Papers*, I, p. 147), the vanishing of six, suitably chosen, second minors of Δ, is sufficient to ensure that they all vanish.

General properties of a quadric given by its equation. After what we have seen we may agree to understand by a quadric the locus determined by the vanishing of any homogeneous quadratic form, $f(x, y, z, t)$, in x, y, z, t. When the contrary is not referred to, we suppose that the determinant Δ does not vanish, so that $f(x, y, z, t)$ can be expressed in the form $A\xi^2 + B\eta^2 + C\zeta^2 + D\tau^2$, where A, B, C, D are rational real functions of the original coefficients, all different from zero, and ξ, η, ζ, τ are linear functions of x, y, z, t, with coefficients which are equally rational real functions of the original coefficients. The coefficients A, B, C, D, if real, may be all positive, or all negative; then, clearly, there is no point, of which the coordinates are all real, for which $f(x, y, z, t)$ vanishes; it is, then, usual, as in the case of a conic (Vol. II, p. 163), to speak of $f(x, y, z, t) = 0$ as representing an imaginary quadric. Or, the coefficients A, B, C, D may consist of two which are positive and two which are negative. In this case, as appears from what we have seen above (p. 7), there are two systems of real lines of which every point is a point of the quadric $f(x, y, z, t) = 0$. Or, finally, three of the coefficients A, B, C, D may be of one sign, and the remaining one of the opposite sign. In this case, the locus $f(x, y, z, t) = 0$ contains real points; for example, if A, B, C be of one sign and D be of the opposite sign, two real points are given by $\eta=0$, $\zeta=0$, $A\xi^2+D\tau^2=0$ But the locus does not contain any real lines. For example, with the particular signs just taken, a real line of the locus would contain a real point lying on the plane represented by $\tau=0$; but $\tau=0$ involves, for points of the locus, $A\xi^2+B\eta^2+C\zeta^2=0$, and, as A, B, C are of the same sign, there is no set of real values of ξ, η, ζ for which this can be satisfied. We see thus that the three cases are distinct, if we take account of the reality of the points of the locus; and, though the reduction to a sum of four squares can be made in an infinite number of ways, yet, when made by rational real substitutions only, as here, the number of positive and negative signs obtained for the coefficients A, B, C, D will be the same, for any particular form $f(x, y, z, t)$ of *real* coefficients, in whatever way the reduction is made.

Certain general properties of the quadric locus may now be enumerated: (1) If a linear transformation be made, from variables x, y, z, t to variables ξ, η, ζ, τ, by equations of the forms

$$\xi=a_1x+b_1y+c_1z+d_1t, \quad \eta=a_2x+b_2y+c_2z+d_2t,$$
$$\zeta=a_3x+\ldots, \quad \tau=a_4x+\ldots,$$

whereby a function $f(x, y, z, t)$ becomes $F(\xi, \eta, \zeta, \tau)$, and if $\xi', \eta', \zeta', \tau'$ be the same linear functions of variables x', y', z', t' as are ξ, η, ζ, τ of x, y, z, t, then

$$\xi'\frac{\partial F}{\partial \xi}+\eta'\frac{\partial F}{\partial \eta}+\zeta'\frac{\partial F}{\partial \zeta}+\tau'\frac{\partial F}{\partial \tau}=x'\frac{\partial f}{\partial x}+y'\frac{\partial f}{\partial y}+z'\frac{\partial f}{\partial z}+t'\frac{\partial f}{\partial t}.$$

For we have
$$\frac{\partial f}{\partial x}=a_1\frac{\partial F}{\partial \xi}+a_2\frac{\partial F}{\partial \eta}+a_3\frac{\partial F}{\partial \zeta}+a_4\frac{\partial F}{\partial \tau},$$

....

$$\frac{\partial f}{\partial t}=d_1\frac{\partial F}{\partial \xi}+d_2\frac{\partial F}{\partial \eta}+d_3\frac{\partial F}{\partial \zeta}+d_4\frac{\partial F}{\partial \tau};$$

if these equations be multiplied in turn by x', y', z', t', and then added, the result follows at once.

It is often convenient to represent the original equations of transformation by

$$(\xi, \eta, \zeta, \tau)=\begin{vmatrix} a_1, & b_1, & c_1, & d_1 \\ a_2, & b_2, & c_2, & d_2 \\ a_3, & b_3, & c_3, & d_3 \\ a_4, & b_4, & c_4, & d_4 \end{vmatrix}(x, y, z, t);$$

the array of coefficients is then called a *matrix*. The matrix obtained from this, by writing as columns what are here written as rows, is then called the *transposed* matrix. Denoting the original matrix by M, and the transposed matrix by $\bar{M}$, we have

$$(\xi, \eta, \zeta, \tau)=M(x, y, z, t),$$

and $$\left(\frac{\partial f}{\partial x}, \frac{\partial f}{\partial y}, \frac{\partial f}{\partial z}, \frac{\partial f}{\partial t}\right) = \overline{M}\left(\frac{\partial F}{\partial \xi}, \frac{\partial F}{\partial \eta}, \frac{\partial F}{\partial \zeta}, \frac{\partial F}{\partial \tau}\right),$$

where the new variables, ξ, η, ζ, τ, occur on the left in the first equation, but the new differential coefficients $\partial F/\partial \xi$, $\partial F/\partial \eta$, etc., occur on the right in the second equation. This is often expressed by saying that the differential coefficients $\partial f/\partial x$, $\partial f/\partial y$, etc., are transformed *contragrediently* from the variables x, y, z, t.

(2) A general quadric, expressed by the equation $f(x, y, z, t) = 0$, is met in two points by the line joining any two given points $(x, y, z,$ and (x', y', z', t'). The necessary and sufficient condition that the two points of meeting should be harmonic conjugates of one another in regard to the given points (x, y, z, t), (x', y', z', t'), is

$$x'\frac{\partial f}{\partial x} + y'\frac{\partial f}{\partial y} + z'\frac{\partial f}{\partial z} + t'\frac{\partial f}{\partial t} = 0;$$

this equation, when $f(x, y, z, t)$ is a quadratic function, is symmetrical in regard to the two points (x, y, z, t), (x', y', z', t'); with the notation originally suggested for the coefficients in $f(x, y, z, t)$, it is (after division by 2)

$$axx' + byy' + czz' + dtt' + f(yz' + y'z) + g(zx' + z'x)$$
$$+ h(xy' + x'y) + u(xt' + x't) + v(yt' + y't) + w(zt' + z't).$$

For the point $(x+\lambda x', y+\lambda y', z+\lambda z', t+\lambda t')$ lies on $f(x, y, z, t) = 0$ if

$$f(x, y, z, t) + \lambda\left(x'\frac{\partial f}{\partial x} + y'\frac{\partial f}{\partial y} + z'\frac{\partial f}{\partial z} + t'\frac{\partial f}{\partial t}\right) + \lambda^2 f(x', y', z', t') = 0;$$

this equation in λ has two roots, and they are equal, and of opposite sign, only if the coefficient of λ vanish.

(3) Particular applications of the result in (2) may be referred to:

It appears that if, from an arbitrary point, (x', y', z', t'), or O, variable lines be drawn to meet the quadric, each in two points, P and Q, the locus of a point, R, which is the harmonic conjugate of O in regard to P and Q, is a plane, with the equation

$$x'\frac{\partial f}{\partial x} + y'\frac{\partial f}{\partial y} + z'\frac{\partial f}{\partial z} + t'\frac{\partial f}{\partial t} = 0.$$

This plane is called the *polar plane* of O in regard to the quadric.

In the particular case in which the quadric is a cone, and all the differential coefficients $\partial f/\partial x$, $\partial f/\partial y$, etc., vanish at the point which is the vertex of this cone, this polar plane passes through the vertex. In the still more particular case when the quadric is a pair of planes, the polar plane passes through the line in which these two planes intersect. Finally, when the quadric is a pair of coincident planes, the polar plane coincides with these.

It appears, also, that if (x', y', z', t') be a point of the quadric $f(x, y, z, t) = 0$, which we now suppose not to be a cone or a particular case of this, so that one of the two roots of the above quadratic equation for λ^{-1} is zero, and if (x, y, z, t) be taken on the polar plane of (x', y', z', t'), but not on $f(x, y, z, t) = 0$, then the second value of λ^{-1} is also zero, so that the two points, where the line joining the points (x', y', z', t'), (x, y, z, t) meets the quadric, both coincide with (x', y', z', t'); while if, with (x', y', z', t') on the quadric, the point (x, y, z, t) be both on the polar plane of (x', y', z', t'), and also such that $f(x, y, z, t) = 0$, then the quadratic equation for λ is satisfied for every value of λ, namely every point of the line joining (x', y', z', t') to (x, y, z, t) lies on the quadric. In general the intersection of the quadric surface with a plane is a conic, being a locus met by a line of the plane in two points; in the case spoken of, when the plane is the polar plane of a point of the quadric, this conic, we see, contains a line, and, therefore, also another line, as part of itself, and consists of these two lines. Thus for a general quadric, the polar plane of any point lying on the quadric, which for such a point is called the *tangent plane* of the quadric at the point, meets the quadric in two lines intersecting at the point; any other line, in the tangent plane through this point, has two coincident intersections with the quadric, lying at this point. The point itself is called the *point of contact* of the tangent plane with the quadric. When the quadric considered is a cone, the tangent plane at any point passes through the vertex of the cone, and the two lines in which the tangent plane meets the cone coincide, in the line joining the vertex to the point of contact, as is easy to see; at the vertex itself there is no proper tangent plane. When the quadric considered consists of two planes, the tangent plane at any point, not on the line of intersection of the two planes, coincides with one of these.

(4) The condition that a plane, represented, suppose, by the equation

$$lx + my + nz + pt = 0,$$

should be a tangent plane of the quadric $f(x, y, z, t) = 0$, is that its equation should be capable, for values of x', y', z', t' which satisfy the two equations

$$f(x', y', z', t') = 0, \quad lx' + my' + nz' + pt' = 0,$$

of being written in the form appropriate to that of the tangent plane at (x', y', z', t'), which is

$$x(ax' + hy' + gz' + ut') + \ldots + t(ux' + vy' + wz' + dt') = 0.$$

For this it is necessary and sufficient that the coefficients of x, y, z, t, in the last equation, should be proportional to those in the given equation, namely that there should be a value of σ for which we have the four equations

$$ax' + hy' + gz' + ut' + \sigma l = 0,$$

$$\dots\dots\dots\dots\dots\dots\dots\dots$$

$$ux' + vy' + wz' + dt' + \sigma p = 0,$$

while, also, $$lx' + my' + nz' + pt' = 0.$$

These equations, as we see by adding the first four, after multiplication, respectively, by x', y', z', t', involve $f(x', y', z', t') = 0$. Thus the necessary and sufficient condition, for the given plane to be a tangent plane of the given quadric, is that l, m, n, p should satisfy the equation

$$\begin{vmatrix} a, & h, & g, & u, & l \\ h, & b, & f, & v, & m \\ g, & f, & c, & w, & n \\ u, & v, & w, & d, & p \\ l, & m, & n, & p, & 0 \end{vmatrix} = 0;$$

if the cofactors of a, b, c, etc., in the determinant Δ, of four rows and columns, which is obtained by omitting the last row and column of this determinant, be denoted, respectively, by A, B, C, etc., this condition is

$$Al^2 + Bm^2 + Cn^2 + Dp^2 + 2Fmn + 2Gnl + 2Hlm \\ + 2Ulp + 2Vmp + 2Wnp = 0.$$

As the equation $f(x, y, z, t) = 0$ is the condition that the point (x, y, z, t) should be a point of the quadric, so the condition just obtained is the condition that the plane (l, m, n, p) should be a tangent plane of the quadric. Thus it appears that, as the quadric is the aggregate of points satisfying the most general homogeneous quadratic relation connecting the point coordinates x, y, z, t, so the aggregate of the tangent planes of a quadric is obtained by subjecting the coefficients, l, m, n, p, in the equation of a plane, to the most general homogeneous quadratic relation. For, supposing the determinant Δ not to be zero, we may regard, not the coefficients a, b, c, etc., as fundamental, but the coefficients A, B, C, etc., as fundamental; taking the determinant

$$\begin{vmatrix} A, & H, & G, & U \\ H, & B, & F, & V \\ G, & F, & C, & W \\ U, & V, & W, & D \end{vmatrix}$$

the coefficients a, b, c, etc., each multiplied by Δ^2, are the cofactors of A, B, C, etc., in this last determinant. The case of $\Delta = 0$ will be considered immediately below.

The equation $Al^2 + Bm^2 + \ldots + 2Wnp = 0$, may then be spoken of as the *tangential equation* of the quadric, and l, m, n, p as the *coordinates* of a plane, the whole being analogous to the case of a conic dealt with in Chap. III of Vol. II.

In particular, if $F(l, m, n, p) = 0$ be the tangential equation of a quadric, the point of contact of a particular tangent plane, whose coordinates are l_0, m_0, n_0, p_0, has for its equation

$$l_0 \frac{\partial F}{\partial l} + m_0 \frac{\partial F}{\partial m} + n_0 \frac{\partial F}{\partial n} + p_0 \frac{\partial F}{\partial p} = 0,$$

so that, if $F(l_0, m_0, n_0, p_0)$ be denoted by F_0, the coordinates of the point of contact are $\partial F_0/\partial l_0$, $\partial F_0/\partial m_0$, etc.

When the determinant Δ vanishes, the tangential equation above written, regarded as deduced from the point equation, is not general, being in fact a perfect square. For, by the theorem of reciprocal determinants referred to above, the determinants of two rows and columns such as

$$BC - F^2, \quad AD - U^2, \quad GH - AF, \quad HU - AV,$$

which are equal respectively to $ad - u^2$, $bc - f^2$, $fd - vw$, $fw - vc$, multiplied by Δ, all vanish; thus, when A is not zero, the tangential equation is the same as

$$(Al + Hm + Gn + Up)^2 = 0,$$

and so in other cases.

Taking, however, the general tangential equation, we may shew, as for the point equation above, that it can be written in the form

$$A_1(a_1 l + b_1 m + c_1 n + d_1 p)^2 + \ldots + A_4(a_4 l + b_4 m + c_4 n + d_4 p)^2 = 0.$$

In the most general case none of A_1, A_2, A_3, A_4 vanishes. When one of these vanishes, the planes conditioned by this equation are all those which pass through the tangent lines of a certain conic. When two of the coefficients $A_1, \ldots, A_4$ vanish, the planes conditioned by this equation are all those which pass through one of two particular points. Finally, when only one of the coefficients is present, the planes are all those passing through a particular point. This follows from the principle of duality, the tangential *equation of a point*, of the form

$$x_0 l + y_0 m + z_0 n + t_0 p = 0,$$

being exactly analogous to the equation of a plane. The matter has been explained in detail for a plane, in Chap. III of Vol. II.

(5) Denoting $f(x, y, z, t)$ by $f(x)$, and $f(x', y', z', t')$ by $f(x')$ and

$$\frac{1}{2}\left(x' \frac{\partial f}{\partial x} + y' \frac{\partial f}{\partial y} + z' \frac{\partial f}{\partial z} + t' \frac{\partial f}{\partial t}\right)$$

by $f(x, x')$, we have remarked above that the points

$$(x+\lambda x', \quad y+\lambda y', \quad z+\lambda z', \quad t+\lambda t'),$$

of the line joining (x', y', z', t') to (x, y, z, t), which lie on the quadric $f(x, y, z, t)=0$, are given by the quadratic equation

$$f(x)+2\lambda f(x, x')+\lambda^2 f(x')=0.$$

The line will then meet the quadric in two coincident points if, for an assigned point (x', y', z', t'), the point (x, y, z, t) be such that

$$f(x)f(x')-[f(x, x')]^2=0.$$

Regarded as a condition for (x, y, z, t), this equation represents a quadric. It is, however, evidently satisfied by any point of the line joining (x', y', z', t') to (x, y, z, t). Thus all the lines from (x', y', z', t') which meet the quadric in two coincident points lie on a quadric cone. This is called the *tangent cone*, or the *enveloping cone*, to the given quadric, from the point, and the lines are said to *touch* the quadric. It can in fact be directly verified, (a), that the equation represents a cone; (b), that every one of its tangent planes is equally a tangent plane of the given quadric.

(6) As the general equation $f(x, y, z, t)=0$ contains ten, homogeneously entering, coefficients, and the condition that the quadric should contain an assigned point is a single linear condition for these coefficients, it follows that a quadric can be found containing nine points arbitrarily chosen. For instance, if three non-intersecting lines be given, a quadric can be found containing three points on each of these lines; and then, as each line meets this quadric in more than two points, the line lies entirely upon the quadric. Or again, if three conics be given in space, of which every two have two points in common, by taking another point upon each of the conics, we have nine points in all; the quadric through these nine points will meet the plane of any one of the three given conics in a conic having five points lying thereon, and, therefore, coinciding with it. The quadric thus contains the three given conics.

It has seemed desirable to give the preceding analysis in order to make clear the general conception of a quadric surface; the notion of the equation of a quadric is undoubtedly of great power, and as we have deduced this from the geometrical definition of the surface, it would be pedantic to avoid the use of it. The descriptive theory, however, often gives a better insight into the geometrical figure under consideration. This theory we now resume.

The polar plane of a point in regard to a quadric. Given any quadric, take an arbitrary point, O, not lying thereon. Draw an arbitrary plane through O, meeting this quadric in a conic. Let a line, passing through O, lying in this plane, meet this conic in the points A and B, and let P be the harmonic conjugate of O

in regard to A and B. As the line varies in this plane, passing through O, the locus of P is a line, the polar line of O in regard to this conic. Now let another plane be drawn through O, meeting the former plane in a line l, passing through O, and let the points in which the line l meets the quadric be U and V. If W be the harmonic conjugate of O in regard to U and V, the point W is on the polar lines of O in regard to both the conics, in which the quadric is met by the two planes drawn through O; these two polar lines, therefore, intersect one another. Hence, if a third plane, not containing the line l, be drawn through O, the polar line of O, in regard to the conic in which this third plane cuts the quadric, will intersect the polar lines of O taken in regard to the two former conics, and not pass through the point of intersection of these lines, which is on l. The third polar line will thus lie in the plane of the first two. From this it is clear that, if an arbitrary line be drawn through O, to meet the quadric in the points Q and R, and P be the point which is the harmonic conjugate of O in regard to Q and R, then the locus of P is a plane. This is called the *polar plane* of O in regard to the quadric. From the definition it is manifest that if the polar plane of O pass through P, the polar plane of P passes through O.

The tangent plane of the quadric at any point. The polar plane of O meets the quadric in a conic. Let H be any point of this. Then the second point, H', in which the line OH meets the quadric, coincides with H; for H is the harmonic conjugate of O in regard to H and H'. We therefore say that the line OH *touches* the quadric at the point H. Any plane drawn through OH meets the quadric in a conic; this conic is met by OH in two points which coincide at H, that is, HO is the tangent at H of this conic. Again, the conic, also passing through H, in which the quadric is met by the polar plane of O, has a tangent at H, say t, unless this conic consists of two lines. This line, t, meets the conic in two points which coincide at H, and, therefore, meets the quadric in two points which coincide at H. Consider now the conic in which the quadric is met by the plane containing the two lines HO and t; this conic is met in two points coinciding at H, both by the line HO and by the line t. It cannot, therefore, be a proper conic. but must consist of two lines meeting at H. The plane of these lines is called the *tangent plane* of the quadric at H. It is to be regarded as the polar plane of the point H; it passes through O, and the polar plane of O passes through H. By choosing O suitably, H may be supposed to be any general point of the quadric. But there may be exception if the polar plane of O meets the quadric in two lines, as happens when the quadric degenerates into a cone, or into two planes; the polar plane of O then passes, respectively, through the vertex of the cone, or through the line of intersection of the two planes; in the former case it meets the cone in lines, in the latter case it meets the planes only in their line of intersection; in the former case there is still a tangent plane of the cone, which passes through the vertex, unless H be actually the vertex.

The polar line of any line in regard to a quadric. Let O_1, O_2 be any two points, not lying on the quadric. Their polar planes will meet in a line, say m. Draw through the line O_1O_2 any plane, meeting the line m in a point K. As the polar planes of O_1 and O_2 both pass through K, it follows that the polar plane of K passes through both O_1 and O_2, and therefore contains any point, O_3, of the line O_1O_2. The polar plane of O_3, therefore, passes through K. The lines O_1O_2 and m are thus in reciprocal relation, the polar plane of any point of either of these containing every point of the other, and, therefore, containing the other. Either of these lines is called the polar line of the other in regard to the quadric. A plane drawn through one of these lines will meet the quadric in a conic and meet the polar line in a point; in regard to this conic, this point is the pole of the first line, as follows from the harmonic relation for points on a transversal. The polar planes, in regard to the quadric, of any range of points on one of the lines, form an axial pencil, of which the polar line is the axis, and, as follows from what has just been remarked, and the theory of conics, this axial pencil of planes is related to the range of points.

Let one of these two lines meet the quadric in the two points F and G; then, as the polar plane of any point, O, of the other line, contains the line FG, and hence contains F and G, it follows, as we have seen above, that the tangent plane of the quadric at F, as also the tangent plane at G, both contain O. In other words, as O is any point of the second line, the tangent planes of the quadric at F and G intersect in this line, the polar line of FG. From this it follows, also, that the lines, lying on the quadric, which pass through F, being in the tangent plane of the quadric at F, both meet the polar line of FG. The points of meeting, being points on lines which lie on the quadric, are themselves on the quadric; they are, therefore, the two points in which the quadric is intersected by the polar line of FG. The lines on the quadric which pass through G must, similarly, pass through these two points. Calling these points F' and G', the two lines, on the quadric, which meet at F', determine the tangent plane of the quadric at F', and this, we see, contains FG; similarly the two lines, on the quadric, which meet at G', determine the tangent plane of the quadric at G', and this plane also passes through the line FG. When the quadric is a cone, the lines on the cone, at any point F, coincide in a line which passes through the vertex; the tangent plane of the cone at any point of this line is the same plane as the tangent plane at any other point of this line, and passes through the vertex. The polar line of a line *FG*, where *F*, *G* are points of the cone other than the vertex, is thus a line through the vertex.

Employing a symbolism introduced in connexion with the original definition of a quadric (above, p. 6), we may suppose, in general, that the points *F*, *G* are, respectively, of symbols

$$\lambda A + \mu B + \lambda\mu C + D, \quad \lambda' A + \mu' B + \lambda'\mu' C + D,$$

so that the parameters λ, μ are associated, respectively, with the two lines of the quadric which pass through *F*, and the parameters λ', μ' with those through *G*. The parameters associated with the two lines through *F′* may, therefore, be taken to be (λ, μ'), and those for *G′* to be (λ', μ); thus the symbols for *F′* and *G′* will, respectively, be

$$\lambda A + \mu' B + \lambda\mu' C + D, \quad \lambda' A + \mu B + \lambda'\mu C + D.$$

The coordinates of the points *F*, *G*, *F′*, *G′*, relatively to *A*, *B*, *C*, *D*, will thus be, respectively,

$$(\lambda, \mu, \lambda\mu, 1); \quad (\lambda', \mu', \lambda'\mu', 1); \quad (\lambda, \mu', \lambda\mu', 1); \quad (\lambda', \mu, \lambda'\mu, 1).$$

The coordinates of the three points *F*, *F′*, *G′*, thus, all satisfy the equation

$$x\mu + y\lambda = z + t\lambda\mu;$$

by the theory we have just given this is, therefore, the equation of the tangent plane of the quadric at *F*. If we put

$$x_0 = \lambda, \quad y_0 = \mu, \quad z_0 = \lambda\mu, \quad t_0 = 1,$$

for the coordinates of *F*, this is the same as

$$xy_0 + yx_0 - zt_0 - tz_0 = 0,$$

or
$$\left(x_0 \frac{\partial}{\partial x} + y_0 \frac{\partial}{\partial y} + z_0 \frac{\partial}{\partial z} + t_0 \frac{\partial}{\partial t}\right)(xy - zt) = 0,$$

which is in accordance with preceding results, the equation of the quadric being $xy - zt = 0$.

The polar point, or pole, of a plane in regard to a quadric. There are two senses in which we may very naturally speak of the pole of a plane in regard to a quadric; we shew here that these lead to the same point.

(*a*) Given any plane ϖ, if any three points of this be *A*, *B* and *C*, and we take their polar planes in regard to the quadric, meeting, suppose, in the point *O*; then, as the polar plane of *A* passes through *O*, therefore the polar plane of *O* passes through *A*, and this plane similarly passes through *B* and *C*; it is, thus, the plane ϖ. The polar plane of any other point of the plane ϖ, since this point lies in the polar plane of O, also passes through O. The pole of the plane ϖ, in this sense, is thus the point common to the polar planes of all points of ϖ: and the plane is the polar plane of this point.

(*b*) As, however, we defined the polar plane of a point when the quadric is regarded as a locus of points, so we may define the pole of a plane, in a dual way, when the quadric is generated by an aggregate of planes. When regarded in the former way, the quadric is the aggregate of points lying on all transversals of three non-intersecting lines. We may, however, consider the aggregate of all planes passing through the transversals of three given non-intersecting lines. We have proved above that these planes are in fact the tangent planes of a quadric defined as a locus of points, each plane containing points of the quadric lying on two lines. As we defined the polar plane of a point, O, in regard to the quadric locus of points, by drawing a line through O, and taking the points Q, R of this line which are points of the quadric, and then the locus of the point P of this line which is the harmonic conjugate of O in regard to Q and R, so we may proceed for the aggregate of planes: given a plane ϖ, let l be a line on this plane; through this line l two planes of the aggregate can be drawn, which are in fact tangent planes of the quadric locus; we may take the plane, through the line l, which is the harmonic conjugate of ϖ in regard to these two planes. All these fourth harmonic planes, when l takes different positions on the plane ϖ, have a point, O', in common. This is the point which we might naturally call the pole of the plane ϖ, when the figure is regarded in this dual way. We shew that this point O' is the same as the point O obtained above, of which ϖ is the polar plane.

For the polar line of l, containing all points whose polar planes pass through l, passes through O; and, as we have seen above, this polar line meets the quadric locus in two points, Q and R, whereat the tangent planes of the quadric are such as to pass through l; as the point in which this line meets the plane ϖ is the harmonic conjugate of O, in regard to Q and R, it follows that the plane ϖ is the harmonic conjugate of the plane Ol in regard to the tangent planes Ql and Rl. This being so for every line l of the plane ϖ, the result follows, that O' coincides with O.

Alternative deduction of the polar point of a plane with the help of the symbols. If we are given three arbitrary non-intersecting lines, we have seen that two points of a transversal of these may, by taking suitable points of reference, be supposed to be of coordinates

$$(\lambda, \mu, \lambda\mu, 1), \quad (\lambda, \mu', \lambda\mu', 1),$$

respectively, where λ refers to the particular transversal chosen, and μ, μ' belong to the generators of the opposite system which determine the two points of this. The conditions that a plane, of equation $lx + my + nz + pt = 0$, should contain these two points, are

$$l\lambda + p + \mu(m + \lambda n) = 0, \quad l\lambda + p + \mu'(m + \lambda n) = 0;$$

the conditions that the plane should contain the transversal λ are, therefore,

$$l\lambda + p = 0, \quad m + \lambda n = 0;$$

these are the only conditions for the coordinates (l, m, n, p) of a plane containing a transversal of the three given lines, which may thus be expressed in terms of λ and another variable parameter, σ, in the forms

$$l = -\frac{1}{\lambda}, \quad m = -\frac{1}{\sigma}, \quad n = \frac{1}{\lambda\sigma}, \quad p = 1,$$

which are of the same forms as are the coordinates $(\lambda, \mu, \lambda\mu, 1)$ of a point of the locus. They satisfy the equation $lm - np = 0$; and the pole of an arbitrary plane, $Lx + My + Nz + Pt = 0$, in regard to the aggregate of planes whose coordinates satisfy this relation, calculated in the precisely dual way, will be the point whose equation is

$$\left(L\frac{\partial}{\partial l} + M\frac{\partial}{\partial m} + N\frac{\partial}{\partial n} + P\frac{\partial}{\partial p}\right)(lm - np) = 0,$$

or

$$Lm + Ml - Np - Pn = 0,$$

that is, the point whose coordinates are $(M, L, -P, -N)$. The polar plane of this point in regard to the quadric locus, of which the points satisfy the equation, $xy - zt = 0$, is the plane whose equation is

$$\left(M\frac{\partial}{\partial x} + L\frac{\partial}{\partial y} - P\frac{\partial}{\partial z} - N\frac{\partial}{\partial t}\right)(xy - zt) = 0;$$

this is exactly the original plane $Lx + My + Nz + Pt = 0$.

Examples of some general properties of quadrics. *Examples* 1—19. We may now usefully apply these ideas to deduce some general properties of a quadric.

Ex. 1. The plane, whose equation is

$$-\frac{x}{\lambda} - \frac{y}{\mu} + \frac{z}{\lambda\mu} + t = 0,$$

is at once verified to contain the three points

$$(\lambda, \mu, \lambda\mu, 1), \quad (\lambda, \mu', \lambda\mu', 1), \quad (\lambda', \mu, \lambda'\mu, 1),$$

whatever λ' and μ' may be, that is, it contains the two intersecting generators λ and μ. Correspondingly, if this plane be called $[\lambda, \mu]$, it is at once verified that the point $(\lambda, \mu, \lambda\mu, 1)$ lies on the three planes $[\lambda, \mu]$, $[\lambda, \mu']$, $[\lambda', \mu]$. The plane $[\lambda, \mu]$ is, in fact, the tangent plane of the quadric $xy - zt = 0$ at the point (λ, μ).

Ex. 2. The pole of the plane $Lx + My + Nz + Pt = 0$ in regard to the plane-aggregate which is represented by the general equation $(A, B, C, D, F, G, H, U, V, W \between l, m, n, p)^2 = 0$, has a tangential equation

$$L(Al + Hm + Gn + Up) + \dots + P(Ul + Vm + Wn + Dp) = 0,$$

and this point, therefore, has the coordinates

$$AL + HM + GN + UP, \dots, UL + VM + WN + DP.$$

The polar plane of this point in regard to the locus represented by $(a, b, c, d, f, g, h, u, v, w \between x, y, z, t)^2 = 0$ has the equation

$$(AL + HM + GN + UP)(ax + hy + gz + ut) + \dots$$
$$+ (UL + VM + WN + DP)(ux + vy + wz + dt) = 0.$$

If A, B, C, etc., be the cofactors of a, b, c, etc. in the determinant Δ, as before, this equation can be verified to reduce to

$$\Delta(Lx + My + Nz + Pt) = 0.$$

Ex. 3. Two points are said to be *conjugate* to one another in regard to a quadric when the polar plane of either point contains the other point. Two planes are said to be *conjugate* to one another in regard to a quadric when the pole of either plane lies on the other plane.

Now, supposing the quadric not to be a cone, consider any two planes, α, β, whose line of intersection meets the quadric in the points F and G; let the polar line of FG, which is the line joining the poles, say A and B, respectively, of the planes α, β, meet the quadric in F' and G'. We have seen that the lines, lying on the quadric, which intersect in F, pass through F' and G', as do the lines, lying on the quadric, which intersect in G. If the planes α, β be conjugate to one another, the point A, the pole of the plane α, lies on the plane β, and similarly B on α; so that the planes α, β meet the line $F'G'$, which is the line AB, in points which are harmonic conjugates of one another in regard to F' and G'. The tangent lines at F, of the conics in which the quadric is met by the planes α, β, meet the line $F'G'$, however, in the same two points as do the planes α, β; for these tangent lines lie in the tangent plane of the quadric at F, which, we have seen, contains the generators of the quadric at F, and contains the line $F'G'$. Thus we have the result: *if two conic sections of a quadric be made by planes which are conjugate to one another in regard to the quadric, the tangent lines of these sections, at either one of the two points of the quadric at which the conics intersect, are harmonic conjugates of one another in regard to the two generators of the quadric at this point.*

The fact which corresponds dually to this may be remarked; it will be seen that, incidentally, it has been established by the argument given. To the points of the quadric which lie in a plane, correspond dually the tangent planes of the quadric which pass through a point; as the points lie on a conic, so these planes are the tangent planes of a quadric cone. As the conics in which two planes meet the quadric have two points in common, these being the points of intersection, with the quadric, of the line of intersection of the two planes, so the tangent cones drawn to the quadric, from two points, have two tangent planes in common, these being the tangent planes drawn to the quadric from the line joining the vertices of the two cones. To the tangent line of a plane conic section of the quadric, at any point of this, corresponds a generator of a tangent cone. To the tangent lines of two plane sections, at a common point, which both lie in the tangent plane of the quadric at this point, correspond, then, two generators of two cones of contact of the quadric, lying in a common tangent plane of these cones, and intersecting in the point where this plane touches the quadric. If A and B be the vertices of these cones, and F be one of the two points of the quadric for which the tangent plane contains the line AB, this plane is one of the tangent planes of the quadric which pass through A, and, therefore, a tangent plane of the cone of contact whose vertex is A; and is, similarly, a tangent plane of the cone of vertex B. The lines AF, BF are the two generators of the cones which correspond to the tangents at a common point of the plane sections. By what we have seen above, if A and B be conjugate points in regard to the quadric, so that they are harmonic conjugates of one another in regard to the two points, F' and G' say, in which the line AB meets the quadric, then, the lines FA, FB are harmonic conjugates in regard to the two generators of the quadric which meet in F.

Ex. 4. We can, in an infinite number of ways, find four points, A, B, C, D, such that each is the pole of the plane containing the other three. Let D be an arbitrary point, and A an arbitrary point lying in the polar plane of D. The polar plane of A will, then, pass through D; and it will meet the polar plane of D in a line. On this line take an arbitrary point, B; the polar plane of B will then pass through A and D, and meet this line in a point C; as C is in the polar plane of D and A and B, its polar plane contains D, A, B. The four points A, B, C, D are, therefore, four such points as were desired.

When referred to A, B, C, D as points of reference, the equation of the quadric must reduce to

$$ax^2 + by^2 + cz^2 + dt^2 = 0,$$

containing only squares of the coordinates. For, if the point D be joined to any point, $xA+yB+zC$, of the plane ABC, the joining line meets the quadric in two points which are harmonic conjugates in regard to D and the point, $xA+yB+zC$, where this line meets the plane ABC. Two such points will be of symbols

$$xA+yB+zC+tD \text{ and } xA+yB+zC-tD;$$

namely, to any point (x, y, z, t) of the quadric is associated a point $(x, y, z, -t)$. Thus the equation of the quadric, $f(x, y, z, t)=0$, must be such as to be the same as $f(x, y, z, -t)=0$, for all values of x, y, z. It is similarly unaffected by change of the sign of x only, or y only, or z only; and thus has the form in question. Or the same result may be obtained by remarking that the polar plane of D, of which the equation is $ux+vy+wz+dt=0$, must reduce to $t=0$, so that $u=v=w=0$; and similarly for A, B, C.

It follows that if $a_r x+b_r y+c_r z+d_r t=0$ be the equations, for $r=1, 2, 3, 4$, of four planes such that any three intersect in the pole of the remaining one, the equation of the quadric is capable of the form

$$\sum_{r=1}^{4} A_r(a_r x+b_r y+c_r z+d_r t)^2=0.$$

Further, the tangential equation of the quadric

$$ax^2+by^2+cz^2+dt^2=0 \text{ is } a^{-1}l^2+b^{-1}m^2+c^{-1}n^2+d^{-1}p^2=0.$$

And, if (x_r, y_r, z_r, t_r) be four points, for $r=1, 2, 3, 4$, such that any one is the pole of the plane containing the other three, the tangential equation of the quadric is capable of the form

$$\sum_{r=1}^{4}(lx_r+my_r+nz_r+pt_r)^2=0.$$

For a cone, one of the four points, say D, must be at the vertex; the equation then is $ax^2+by^2+cz^2=0$; the polar plane of *any* point of the line DA is then the plane DBC, and so for DB and DC; the plane $t=0$ is then arbitrary.

Four points A, B, C, D, of which each is the pole of the plane containing the other three, are said to form a self-polar tetrad in regard to the quadric.

Ex. 5. When two lines, l, m, are such that the polar line, l', of l, in regard to a quadric, intersects the other line m, it can be shewn that the polar line, m', of m intersects l. For, if M be the point where l' meets m, the polar plane of M, a point of l', contains the line l; but as M is on m, this polar plane also contains the line m'. Thus l and m' meet. Two such lines, l, m, are said to be *conjugate* in regard to the quadric.

It can be shewn that, if A, B, C, D be four points such that two pairs of opposite joining lines of these be conjugate, say CA conjugate to BD and AB conjugate to CD, then also the third pair of joins, BC and AD, are conjugate lines. Let D', A', B', C' be the poles, respectively, of the planes ABC, BCD, CAD, ABD, so

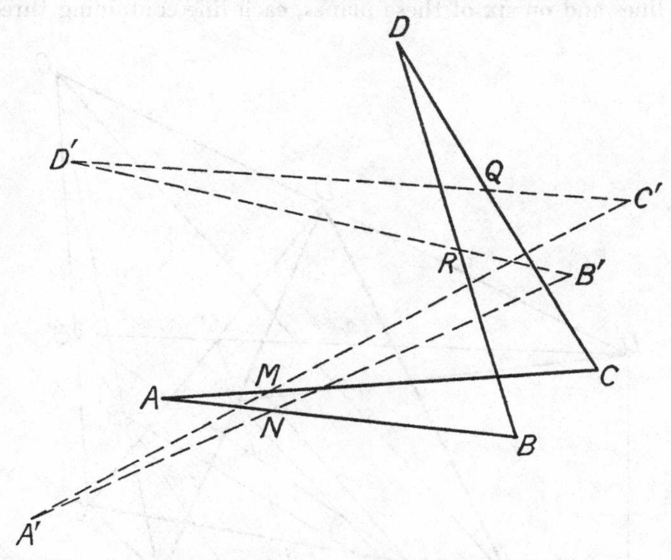

that $C'D'$, which is the polar line of AB, intersects CD, say in Q; and similarly, $B'D'$, $C'A'$, $B'A'$, the respective polar lines of AC, BD, CD, meet the joins BD, AC, AB, say in the points R, M, N.

Considering the section, of the quadric and of the planes DBC, DCA, DAB, by the plane $A'B'C'$, we obtain a conic, and three lines, which, in virtue of the definition of a polar plane by a harmonic range, are the polar lines, with respect to this conic, respectively of the points A', B', C'. The three intersections of these lines respectively with $B'C'$, $C'A'$, $A'B'$ are thus in line (Vol. II, p. 31). Thus, the line $B'C'$ intersects the line MN in a point lying on the plane DBC, which is, therefore, on the line BC. By considering the section by the plane $D'B'C'$, we similarly prove that QR and $B'C'$ intersect on BC. Let this point of intersection of $B'C'$, QR, MN, BC be called L. A similar argument will shew that QM, RN, AD and $A'D'$ meet in a point, say P. It is then easy to prove that any two of the four lines AA', BB', CC', DD' meet one another, so that these four lines all meet in one point, say O, and the two tetrads A, B, C, D and A', B', C', D' are in perspective from O, corresponding lines and planes, belonging to these, intersecting the plane $QMNR$ in the same point or line. The consequence that BC, being met by $B'C'$, is conjugate to AD, is contained in this statement.

The complete figure so obtained is one in which there are fifteen points, twenty lines and fifteen planes, each point lying on four of these lines and on six of these planes, each line containing three of

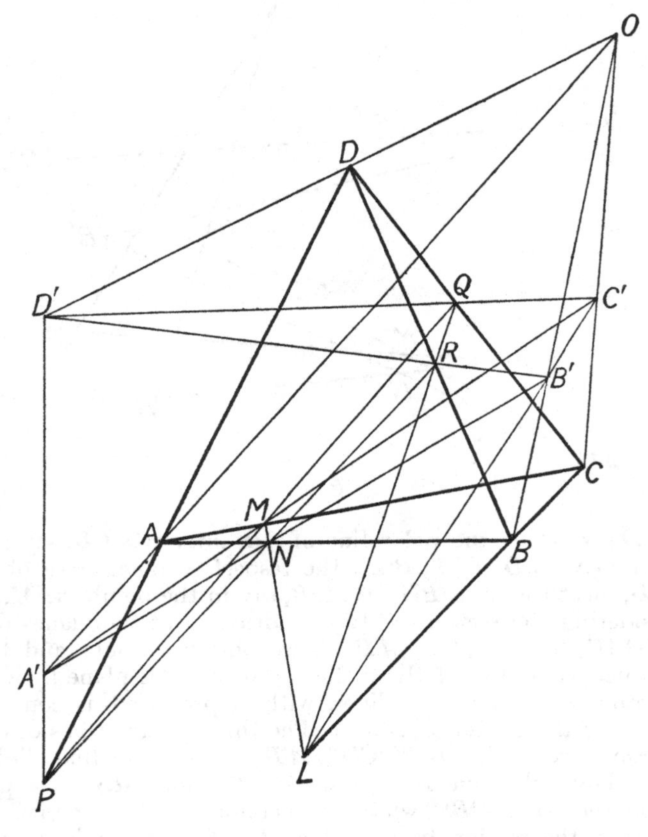

the points and lying in three of the planes, each plane containing six of the points, and four of the lines. So that the structure of the figure may be indicated by the scheme

$$15\,(\,\cdot\,, 4, 6)\;20\,(3, \cdot\,, 3)\;15\,(6, 4, \cdot\,);$$

the figure is in fact the same as that which is used in the proof of Desargues' theorem for two triangles lying in the same plane (of which a diagram is given as frontispiece to Vol. I).

In this figure the points A', B', C', D' are introduced as the poles of certain planes, and these are in perspective with A, B, C, D

from the point O. It can however be shewn that the points of the figure are symmetrical in their relations to the quadric. For first, the line AA', joining the poles of the planes $B'C'D'$, BCD, is the polar line of the intersection of these, that is, of the line QR; and, similarly, the lines BB', CC', DD' are, respectively, the polars of the lines MQ, NR, MN, so that the point O is the pole of the plane, $MQRN$, containing these four lines; and also, the point P, where MQ, NR intersect, is the pole of the plane OBC, containing the lines BB', CC'; and, similarly, L, M, N, Q, R are, respectively, the poles of the planes OAD, OBD, OCD, OAB, OCA. Thus the six points in which the polar plane of O meets the joins of A, B, C, D (which are also the points in which this plane meets the joins of A', B', C', D'), are the poles of the planes joining O to the respectively opposite joins of A, B, C, D (which are the same as the planes joining O to the corresponding joins of A', B', C', D'). Every one of the fifteen points of the figure is, however, the centre of perspective of two tetrads of points, of which the respective sets of six joins occur in the figure; the remaining six points of the figure are then on the polar plane of the centre of perspective, and are the poles of the six planes of the figure passing through this centre. For each of the two tetrads, a join, l', is conjugate to the opposite join, l, the polar line of l being a join in the second tetrad, meeting l'. Either tetrad consists of the poles of the planes containing three points of the other tetrad. The point O and A, B, C, D together form a pentad of points, with the property that the pole of the plane containing any three of these lies on the line joining the other two; the same statement is true of the pentad O, A', B', C', D'. For each of the fifteen points of the figure a similar pair of pentads exists. Such a pentad may be called a *self-conjugate pentad*. It may be regarded as a generalisation of a self-polar tetrad, just as, in a plane, a set of four points, of which every pair of opposite joins consists of conjugate lines in regard to a conic (Vol. II, p. 33), may be regarded as a generalisation of a self-polar triad.

Conversely given two tetrads of points which are in perspective, there is a definite quadric in regard to which the completed figure has the reciprocal properties which have been announced.

The whole may be discussed with the symbols; and it appears that the quadric can be expressed, in point coordinates, or in tangential coordinates, by an equation involving five squares, for each of the ways in which the figure may be regarded. Let A, B, C, D be four points such that the lines AC, BD are conjugate, and, also, the lines AB, CD are conjugate, in regard to a certain quadric surface. Let the equation of this surface, in point coordinates referred to A, B, C, D, be $(a, b, c, d, f, g, h, u, v, w \between x, y, z, t)^2 = 0$, and A', B', C', D' be, as before, the poles of the planes BCD, CAD, ABD, ABC; then the planes $A'B'D'$, $B'C'D'$, which are the polar planes of C and A respectively, have the respective equations

$$gx + fy + cz + wt = 0, \quad ax + hy + gz + ut = 0;$$

in order that $B'D'$, the line of intersection of these planes, should intersect BD, the points, given by $fy + wt = 0$, $hy + ut = 0$, where these planes meet BD, must be the same, so that $fu = hw$. In order that $C'D'$ may intersect CD, the results of putting $x = 0$, $y = 0$ in the equations $ax + hy + gz + ut = 0$, $hx + by + fz + vt = 0$, namely $gz + ut = 0$, $fz + vt = 0$, must be the same, so that $fu = gv$. From $gv = hw$ it then follows at once that the joins BC, AD are conjugate. It follows also, writing σ in place of any one of the equal expressions vw/f, wu/g, uv/h, that the equation of the quadric is capable of the form

$$(a - \sigma^{-1}u^2)\,x^2 + (b - \sigma^{-1}v^2)\,y^2 + (c - \sigma^{-1}w^2)\,z^2 + (d - \sigma)\,t^2 + \sigma^{-1}(ux + vy + wz + \sigma t)^2 = 0;$$

here $x = 0$, $y = 0$, $z = 0$, $t = 0$ are the equations of the planes BCD, ..., ABC; and it can be shewn that $ux + vy + wz + \sigma t = 0$ is the equation of the plane containing the six points of intersection of corresponding joins of the two tetrads A, B, C, D and A', B', C', D'. This plane is the polar of the point, O, in which the lines AA', BB', CC', DD' meet. It will appear that, if the cofactors of $a, b, c, \ldots$ in the determinant Δ, formed by the coefficients in the equation of the quadric, be, as above (p. 24), denoted by $A, B, C, \ldots$, then there exist also the relations $FU = GV = HW$, and that, if λ denote any one of the equal expressions VW/F, WU/G, UV/H, then the coordinates of O are (U, V, W, λ). That the polar plane of this point should have the equation above given involves identities which are reducible to those given.

We may investigate the theorem with the tangential equation of the quadric, supposed of the form

$$(A, B, C, D, F, G, H, U, V, W \between l, m, n, p)^2 = 0.$$

The condition that the line $B'D'$ should meet the line BD is that a plane with equation of the form $x = mz$ should contain the points B', D', whose coordinates are, respectively, (H, B, F, V) and (U, V, W, D); this requires that $HW = FU$. The condition that the line $C'D'$ should meet the line CD, similarly, requires that the points (G, F, C, W), (U, V, W, D) should lie in a plane with equation of the form $x = ny$; and this involves that $GV = FU$. The consequence $GV = HW$, of these equations, is easily seen to secure that the joins BC, AD are conjugate in regard to the quadric. Denoting the equal quantities VWF^{-1}, etc., as just said, by λ, these identities allow the tangential equation of the quadric to be put into the form

$$(A-\lambda^{-1}U)l^2+(B-\lambda^{-1}V)m^2+(C-\lambda^{-1}W)n^2+(D-\lambda)p^2 + \lambda^{-1}(Ul+Vm+Wn+\lambda p)^2=0,$$

where $l=0$, $m=0$, $n=0$, $p=0$, $Ul+Vm+Wn+\lambda p=0$, are, respectively, the tangential equations of the points A, B, C, D, O.

In general, we may determine the tangential equation of a quadric whose point equation is

$$Ex^2+Qy^2+Rz^2+Kt^2+(ex+qy+rz+kt)^2=0.$$

For this, we have to express that, for a point (x', y', z', t') satisfying this equation, and the equation $lx'+my'+nz'+pt'=0$, the equation

$$Exx'+Qyy'+Rzz'+Ktt'+(ex'+qy'+rz'+kt')(ex+qy+rz+kt)=0$$

is the same as $lx+my+nz+pt=0$; namely that, with M' written for $ex'+qy'+rz'+kt'$, we have

$$\frac{Ex'+M'e}{l}=\frac{Qy'+M'q}{m}=\frac{Rz'+M'r}{n}=\frac{Kt'+M'k}{p}.$$

These equations, if $lx'+my'+nz'+pt'=0$, involve that (x', y', z', t') lies on the quadric. Denoting these equal fractions by θ, we have

$$Ex'+M'e-\theta l=0, \quad \ldots, \quad Kt'+M'k-\theta p=0,$$

from which we infer

$$M'+M'(e^2E^{-1}+\ldots+k^2K^{-1})-\theta(elE^{-1}+\ldots+kpK^{-1})=0;$$

if $1+e^2E^{-1}+\ldots+k^2K^{-1}$ be denoted by ω, and $elE^{-1}+\ldots+kpK^{-1}$ by L, this equation is $M'=\omega^{-1}\theta L$. We, therefore, have

$$Ex'+\theta(e\omega^{-1}L-l)=0, \ldots, Kt'+\theta(k\omega^{-1}L-p)=0,$$

so that, by $lx'+\ldots+pt'=0$, we deduce

$$E^{-1}l^2+\ldots+K^{-1}p^2-\omega^{-1}L^2=0;$$

this is the tangential equation sought. More explicitly it is

$$\left(1+\frac{e^2}{E}+\ldots+\frac{k^2}{K}\right)\left(\frac{l^2}{E}+\ldots+\frac{p^2}{K}\right)-\left(\frac{e}{E}l+\ldots+\frac{k}{K}p\right)^2=0.$$

It is easily seen that a special form of self-conjugate pentad is constituted by a self-polar tetrad and an arbitrary point.

A particular case of a tetrad, of which each pair of opposite joins are conjugate lines in regard to a quadric, arises when the quadric, regarded as the aggregate of its tangent planes, reduces to all the planes passing through the tangent lines of a conic. This degeneration of a quadric is that which is the dual of the case when the quadric, regarded as the locus of its points, reduces to all the points on the generating lines of a quadric cone. In the latter case, the polar plane of any point passes through the vertex of the cone. In the former case, which we consider here, the pole of any plane is a point in the plane of the conic, being the pole, in regard to the conic, of the line in which the plane meets the plane of the conic; the polar line of any line, which does not lie in the plane of the conic, is the line, in this plane, which is the polar line, in regard to the conic, of the point in which the former line meets this plane. Two lines, of which neither lies in the plane of the conic, are then conjugate lines if they meet in this plane in two points which are conjugate points in regard to this conic. This being understood, we still have the result that, if the joins AC, BD, of four points A, B, C, D, be conjugate lines in regard to this degenerate quadric, and also the joins AB, CD be conjugate lines, then the remaining pair of joins BC, AD are conjugate lines. And we have the further result that the lines joining A, B, C, D, respectively to the poles A', B', C', D', of the planes BCD, CAD, ABD, ABC, all of which are in the plane of the conic, are four lines which meet in a point. In this case the points L, M, N, P, Q, R, are the points where the joins of A, B, C, D meet the plane of the conic.

Ex. 6. The unsymmetrical algebra of Ex. 5 may be replaced by symmetrical algebra, which suggests itself naturally if we regard the figure as arising by projection of a figure in four dimensions. The theorem so suggested is one for which there is an analogue in space of any number of dimensions; for instance we may have, in a plane, a set of four points such that the join of any two is conjugate to the join of the other two, with respect to a conic in that plane. This is analogous to the self-conjugate pentad. By taking the poles of these joins we obtain six other points, lying by threes on four lines. Thereby we have a figure of ten points and ten lines, of which three points lie on each line, and three lines pass through each point. This is analogous to the complete figure obtained above, and is self-polar in regard to the conic; as the figure obtained in Ex. 5 may be regarded as consisting of two tetrads in perspective from a point, together with the six points of intersection of the corresponding joins of the two tetrads, so the figure of which we speak, in a plane, may be regarded as consisting, in ten ways, of two triads in perspective from a point, together with the three points of intersection of corresponding joins of the two triads, these lying in line. The centre of perspective is the pole of this line. The conic in regard to which the polarity has place may have its equation expressed as a sum of four squares, in terms of the three joins of one of the triads which are in perspective, and of the axis of perspective corresponding, and this mode of expression arises in as many ways as the relation in question. But the analytical expression is suggested most naturally by regarding the figure as arising from a figure in three dimensions, that namely of the complete intersection of five planes. This figure is also self-polar, in a particular way, in regard to a quadric cone.

The existence of this polarity may be regarded as a generalisation of the theorem that if we take four arbitrary lines in a plane, and an arbitrary point O of the plane, the pair of lines joining O to two opposite pairs of intersection of the four lines, are harmonic conjugates in regard to a certain pair of lines drawn through O. This is the dual of the Greek theorem that the joins of four points of a plane cut an arbitrary line in pairs of points in involution.

The general theorem to which reference is here made will be understood from the Examples on p. 218 of Vol. II.

Ex. 7. Consider now a tetrad of points in three dimensions, and a fixed quadric, Σ, such that no one of the three pairs of opposite joins of these points consists of lines which are conjugate in regard to Σ. We prove that, if A', B', C', D' be the poles of the planes BCD, CAD, ABD, ABC, respectively, then the four lines, AA', BB', CC', DD', are such that they have an infinite number of transversals, and that they may be regarded as lines belonging to one system of lines of a quadric surface.

The condition that no two opposite joins, such as AB, CD, are conjugate lines, involves that no two of the lines AA', BB', CC', DD' intersect. For, if AA' meets BB', then $A'B'$ meets AB; but $A'B'$ is the polar line of CD, and only meets AB when AB, CD are conjugate.

Now consider the plane $A'B'C'$; let it meet the lines DA, DB, DC respectively in A_1, B_1, C_1. Then, in virtue of the definition of the polar plane of a point in regard to a quadric, by means of ranges of four harmonic points, it follows that the lines B_1C_1, C_1A_1, A_1B_1 are the polar lines, respectively of A', B', C', in regard to the conic in which the plane $A'B'C'$ meets the quadric Σ. Thus, by a theorem for two triads which arise by taking polars in regard to a conic (Vol. II, p. 31), the triads A', B', C' and A_1, B_1, C_1 are in perspective, and the lines $A'A_1$, $B'B_1$, $C'C_1$ meet in a point. Thus the planes DAA', DBB', DCC' meet in a line; and this line meets AA', BB' and CC'; it evidently also meets DD', at D. By a similar argument there is a line through each of the points A, B, C which meets all the lines AA', BB', CC', DD'. The three lines AA', BB', CC', of which, as we have seen, no two intersect, determine a quadric surface, containing all the points of all the transversals of these three lines; the line DD' contains four points of this quadric surface, namely the points where it is met by the four transversals of AA', BB', CC', DD' which are found above. Thus DD' lies entirely on this quadric surface.

Let the tangential equation of the given quadric surface, Σ, referred to the points A, B, C, D, be, as before,

$$(A, B, C, D, F, G, H, U, V, W \between l, m, n, p)^2 = 0.$$

Then, as the poles of the planes ABC, BCD, etc., are (U, V, W, D), (A, H, G, U), etc., the planes DAA', DBB', DCC' are at once seen to intersect in the line, through the point D, whose equations are $Fx = Gy = Hz$; and the planes ABB', ACC', ADD' similarly intersect in the line whose equations are $V^{-1}y = W^{-1}z = FV^{-1}W^{-1}t$. These two lines are at once verified to lie entirely upon the quadric surface whose equation is

$$(GV - HW)(Fxt + Uyz) + (HW - FU)(Gyt + Vzx) + (FU - GV)(Hzt + Wxy) = 0,$$

and, as this equation is symmetrical, it is that containing the four lines AA', BB', CC', DD'. This equation is independent of the coefficients A, B, C, D.

Let this quadric be denoted by $\Phi(A, B, C, D)$, and let E be any point lying thereon. We may similarly form, with poles of the planes ABC, BCE, CAE, ABE, taken in regard to the same quadric Σ, a quadric $\Phi(A, B, C, E)$. One common point of these quadrics is the point D', which is the pole of the plane ABC in regard to the quadric Σ, and the lines $D'D$, $D'E$ lie, by construction, respectively on the quadrics $\Phi(A, B, C, D)$ and $\Phi(A, B, C, E)$; through D' there passes another generator of each of these quadrics. This other generator is, in fact, the same for both. For, if P, Q, R be the poles, in regard to Σ, respectively of the planes BCE, CAE, ABE, the three points D', A', P, being poles of planes which intersect in the line BC, are on a line, the polar line of BC in regard to Σ; therefore the plane $D'AA'$, containing $D'A'$, equally contains $D'P$. The generator of $\Phi(A, B, C, D)$ of which we are speaking is a line through D' meeting the lines AA', BB', CC'; thus the planes $D'AA'$, $D'BB'$, $D'CC'$ meet in a line, and, of these, the plane $D'AA'$ contains $D'P$, as we have seen; this common line, then, equally meets the lines AP, BQ, CR, and is the generator spoken of, drawn through D', of the quadric $\Phi(A, B, C, E)$.

It can be further shewn that the quadric $\Phi(A, B, C, E)$ passes through the point D. But this result follows from the less particular result given in the following example (8).

Ex. 8. Let A, B, C and P, Q, R be any six points, in three dimensions; suppose that the polar lines of the lines QR, RP, PQ, in regard to any quadric, Σ, which will be lines passing through the pole, say O, of the plane PQR in regard to the quadric Σ, meet the plane ABC in points A', B', C', respectively, such that the lines AA', BB', CC' meet in a point. Then, if the polar lines, in regard to Σ, of the lines BC, CA, AB, which will be lines passing through the pole, say D, of the plane ABC, meet the plane PQR, respectively, in points P', Q', R', the lines PP', QQ', RR' also meet in a point.

For, by the hypothesis made, the planes $AA'O$, $BB'O$, $CC'O$ meet in a line. Now A is the intersection of the lines AB, AC, and is thus the pole of the plane containing the polar lines of these, DR', DQ', respectively, namely, A is the pole of the plane $DQ'R'$; and $A'O$ is the polar line of QR. Thus, the plane joining A to the line $A'O$ is the polar of the point in which the plane $DQ'R'$ is met by the line QR; in other words, the plane $AA'O$ is the polar of the point of intersection of the lines $Q'R'$ and QR. Similarly the planes $BB'O$ and $CC'O$ are, respectively, the polar planes of the points $(R'P', RP)$ and $(P'Q', PQ)$. As these three planes meet in a line, it follows that the points $(Q'R', QR)$, $(R'P', RP)$, $(P'Q', PQ)$ are in line, so that the triads P', Q', R' and P, Q, R are in perspective, as was to be shewn.

The proposition dual to the above may also be enunciated; but first it may be well to remark that three planes, α, β, γ, are said to be in perspective with three planes, α', β', γ', whose point of intersection is the same as the point of intersection of α, β, γ, when the plane containing the lines $\beta\gamma$, $\beta'\gamma'$, the plane containing the lines $\gamma\alpha$, $\gamma'\alpha'$, and the plane containing the lines $\alpha\beta$, $\alpha'\beta'$, are three planes which meet in a line. When this is so, the line of intersection of the planes α, α', the line β, β' and the line γ, γ', are three lines which lie in a plane. This being clear, the dual of the result above given is this: let α, β, γ be three planes, meeting in a point (α, β, γ), and (λ, μ, ν) be three other planes, meeting in a point (λ, μ, ν). Consider the polar lines of the lines (μ, ν), (ν, λ) and (λ, μ); these polars are lines lying in the polar plane of the point (λ, μ, ν). Suppose that the planes joining the point (α, β, γ) to these polar lines are in perspective with the planes α, β, γ. Then, also, the polar lines of the lines (β, γ), (γ, α), (α, β)—which polars are lines lying in the polar plane of the point (α, β, γ)—when joined by planes to the point (λ, μ, ν), give three planes in perspective with the planes λ, μ, ν.

A particular case of this is the result referred to in the preceding example. Let DBC, DCA, DAB, or λ, μ, ν, be three planes, whose respective poles are A', B', C', with reference to a quadric Σ. Let E be a point from which can be drawn a line to meet the lines AA', BB', CC', so that the planes EAA', EBB', ECC' meet in a line. From this it follows that if the planes EBC, ECA, EAB be called, respectively, α, β, γ, the planes joining E to the polar lines, respectively $B'C'$, $C'A'$, $A'B'$, of λ, μ, ν, are in perspective with the planes α, β, γ. Now let P, Q, R, respectively, be the poles of the planes EBC, ECA, EAB, so that the planes joining D to the polars of the lines EA, EB, EC are, respectively, the planes DQR, DRP, DPQ. By what we have seen above, these planes are in perspective with the planes λ, μ, ν, respectively. Therefore the planes DAP, DBQ, DCR meet in a line. Hence, a line can be drawn from D to meet the lines AP, BQ, CR. That is, as stated in the last example, when E is a point of the quadric $\Phi(A, B, C, D)$, then D is a point of the quadric $\Phi(A, B, C, E)$.

Ex. 9. It has been remarked, in Ex. 5, that when two tetrads of points, in three dimensions, are in perspective, there is a definite quadric, in regard to which each point of either tetrad is the pole of the plane containing three points of the other tetrad. A particular case is that when three of the points of either tetrad lie upon their corresponding planes, which are thus tangent planes of the quadric. This case arises when we consider a skew hexagon, formed with three generators of a quadric of one system, and three generators of the other system (Dandelin, *Gergonne's Annales de Math.* XV (1824, 5), p. 387; Hesse, *Crelle*, XXIV (1842), p. 40); and the figure leads to a proof of the concurrence, in a Steiner point, of three of the Pascal lines of six points of a conic, and to the theorem that two of these Steiner points are conjugate points in regard to the conic (Vol. II, Note II, p. 231).

It will be simplest, first, to describe the figure which is obtained: let O, A, B, C be four points of which no three are in line; let A', B', C' be any points lying respectively on OA, OB, OC, the lines BC', $B'C$ meeting in D, the lines CA', $C'A$ meeting in E, and the lines AB', $A'B$ meeting in F. The lines BC, EF then meet on $B'C'$ at the point, L, where the planes $OB'C'$, $A'BC$, AEF meet; so CA, FD, $C'A'$ meet, say in M, and the lines AB, DE, $A'B'$ meet, say in N; the points L, M, N are in the line in which the planes ABC, $A'B'C'$, DEF meet. Further, the lines AD, BE, CF meet in a point, say X; also the lines $A'D$, $B'E$, $C'F$ meet in a point, say X'; and the points O, X, X' are in line. The two tetrads A, B, C, X and A', B', C', X' are, thus, in perspective from O, the plane of perspective being D, E, F, L, M, N; but A' lies on the plane BCX, as A lies on the plane $B'C'X'$, and, similarly, B, B', C, C' lie each on their corresponding planes; the figure is thus a particular case of one often considered, of which the scheme of the indices of structure is $15(\cdot, 4, 6)\ 20(3, \cdot, 3)\ 15(6, 4, \cdot)$. If we take, for the symbols of A, B, C, in regard to A', B', C', O, $A = O + A'$, etc., we have $D = O + B' + C'$, etc.; $X = 2O + A' + B' + C'$; $X' = O + A' + B' + C'$.

We may initiate the figure by taking six generators, AB', $B'C$, CA', $A'B$, BC', $C'A$, which we shall call, respectively, p, q', r, p', q, r', of a quadric surface. The point A, where the generators AB', AC' intersect, is then the pole of the plane $B'C'X'$, and so on; the line

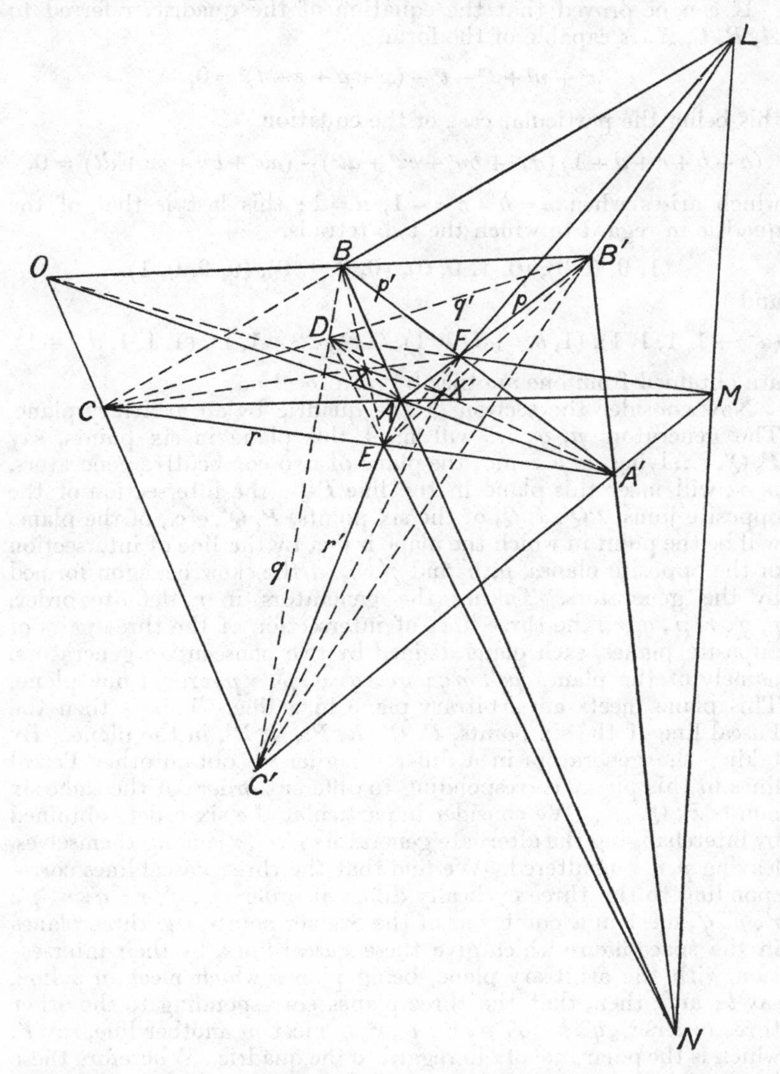

AD is the polar line of the line $B'C'$, etc., and X is the pole of the plane $A'B'C'$; so the line $A'D$ is the polar line of the line BC, etc., and X' is the pole of the plane ABC; the line AA' is the polar

line of the line EF, etc.; hence the line AXX' is the polar line of the line LMN.

It can be proved that the equation of the quadric, referred to A, B, C, X, is capable of the form

$$x^2 + y^2 + z^2 - t^2 - (x + y + z - t)^2 = 0,$$

this being the particular case of the equation

$$(a + b + c + d + 1)(ax^2 + by^2 + cz^2 + dt^2) - (ax + by + cz + dt)^2 = 0,$$

which arises when $a = b = c = -1$, $d = 1$; this last is that of the quadric in regard to which the two tetrads,

$$(1, 0, 0, 0), (0, 1, 0, 0), (0, 0, 1, 0), (0, 0, 0, 1),$$

and

$$(a^{-1} + 1, 1, 1, 1), (1, b^{-1} + 1, 1, 1), (1, 1, c^{-1} + 1, 1), (1, 1, 1, d^{-1} + 1),$$

are obtained from one another by reciprocation.

Now consider the section of the quadric by an arbitrary plane. The generators $p, q', \ldots$ will meet this plane in six points, say $P, Q', \ldots$, lying on a conic; the plane of two consecutive generators, p, q', will meet this plane in the line PQ'; the intersection of the opposite joins, PQ', $P'Q$, of the six points, P, Q', etc., of the plane, will be the point in which the plane is met by the line of intersection of the opposite planes, p, q' and p', q, of the skew hexagon formed by the generators. Taking the generators in a definite order, p, q', r, p', q, r', the three lines of intersection of the three pairs of opposite planes, each plane defined by two consecutive generators, namely of the planes pq', $p'q$; qr', $q'r$; rp', $r'p$, are in one plane. This plane meets an arbitrary plane in a line. This is then the Pascal line of the six points, P, Q', R, P', Q, R', in the plane. By taking the generators in a different order we obtain other Pascal lines in this plane, corresponding to different orders of the same six points $P, Q', \ldots$. We consider in particular the six orders obtained by interchanging the alternate generators p', q', r' among themselves, leaving p, q, r unaltered. We find that the three Pascal lines corresponding to the three cyclically different orders p', q', r'; q', r', p'; r', p', q', meet in a point, one of the Steiner points, the three planes in the space figure which give these Pascal lines, by their intersection with the arbitrary plane, being planes which meet in a line, say l; and, then, that the three planes corresponding to the other three orders r', q', p'; q', p', r'; p', r', q' meet in another line, say l', which is the polar line of l in regard to the quadric. Wherefore these latter give rise to a Steiner point, for the conic, which is conjugate to the former Steiner point in regard thereto. The details may be tabulated thus:

the orders of the generators

$$p, p', r, r', q, q'; \quad p, r', r, q', q, p'; \quad p, q', r, p', q, r';$$
$$p, q', r, r', q, p'; \quad p, p', r, q', q, r'; \quad p, r', r, p', q, q',$$

correspond respectively to the skew hexagons

$$B', F, A', E, C', D; \quad F, A, E, C, D, B; \quad A, B', C, A', B, C';$$
$$F, B', C, E, C', B; \quad A, F, A', C, D, C'; \quad B', A, E, A', B, D;$$

the pairs of opposite planes of these, each plane containing two consecutive sides of the skew hexagon, meet on three lines which are the joins, respectively, of the three points

$$A, B, C; \quad A', B', C'; \quad D, E, F; \quad A, A', D; \quad B, B', E; \quad C, C', F;$$

the planes determined by the first three triads of points meet in the line LMN; the planes determined by the second three triads of points meet in the line OXX'. We have seen that this line is the polar line of the line LMN.

This proves the statement made.

Ex. 10. We have considered above, in Ex. 5, a set of five points which are such that the pole of the plane containing any three of them, taken in regard to a quadric, lies on the join of the other two, calling it a self-conjugate pentad. We consider now a set of six points which are such that the pole of the plane containing any three of them, taken in regard to a quadric, lies on the plane containing the other three. The pentad was such that the tetrad, constituted by any four of its points, had each of its three pairs of opposite joins as conjugate lines, in regard to the quadric. For a hexad, we take any four arbitrary points A, B, C, D, and consider the quadric, $\Phi(A, B, C, D)$, of which four generators are the lines joining A, B, C, D, respectively, to the poles, A', B', C', D', of the planes BCD, CAD, ABD, ABC, taken in regard to the fixed quadric, Σ. If E and F form, with A, B, C, D, such a hexad as is desired, the plane EFD must contain the pole, D, of the plane ABC, so that the line EF must meet DD'; and it must similarly meet AA', BB', CC'. Thus EF must be a generator of $\Phi(A, B, C, D)$, of the opposite system from the lines AA', etc. Again, the plane ADF must contain the pole, P, of the plane BCE, or DF must meet AP; similarly, DF must meet BQ and CR, where Q and R are, respectively, the poles of the planes CAE and ABE. Thus DF must be a generator of the quadric $\Phi(A, B, C, E)$, of the opposite system from the lines AP, etc. We therefore proceed as follows: we take E to be any point of $\Phi(A, B, C, D)$, and on the generator of this through E, of the opposite system from the lines AA', etc., we take the point F, in which the quadric $\Phi(A, B, C, E)$, which passes through E, meets this generator again. Then as EF meets DD', the line ED', which lies on $\Phi(A, B, C, E)$, meets DF; we have shewn (above, Ex. 8) that $\Phi(A, B, C, E)$ contains D; thus the line DF contains three points of $\Phi(A, B, C, E)$, and is a generator of this, through D, meeting ED', and, thus, of the opposite system from the lines AP, etc. We can now shew that the hexad A, B, C, D, E, F, so constructed, is such a self-conjugate hexad as is desired, in regard to the quadric Σ.

For first, as EF meets DD', the plane DFF' contains the pole, D', of the plane ABC; for a similar reason, the planes AEF, BEF, CEF, are, respectively, conjugate to the opposite planes of the hexad, BCD, CAD, ABD. Again, as DF is a generator of $\Phi(A, B, C, E)$, of the opposite system from the lines AP, etc., it follows that AP meets DF, and the plane ADF contains the pole, P, of the opposite plane, BCE, of the hexad; similarly, the planes, BDF, CDF are conjugate to the respectively opposite planes, CAE, ABE, of the hexad. Finally, we proved (Ex. 8, above), that if E be on $\Phi(A,B,C,D)$, then D is on $\Phi(A, B, C, E)$; by a similar argument, since F is on $\Phi(A, B, C, E)$, it follows that E is on $\Phi(A, B, C, F)$, and the point E may be constructed from F, just as was F from E. Wherefore it follows that DE is a generator of $\Phi(A, B, C, F)$, and that the planes ADE, BDE, CDE are conjugate to the respectively opposite planes, BCF, CAF, ABF, of the hexad. The proof of the property is thus complete, for each of the ten pairs of opposite planes of the hexad. In the particular case when E is so taken that EF contains the pole, D', of the plane ABC, it may be shewn that A, B, C, E, F constitute a self-conjugate pentad.

It has been remarked that the equation of the quadric $\Phi(A,B,C,D)$, referred to A, B, C, D, if the tangential equation of the fixed quadric, Σ, be $(A, B, C, D, F, G, H, U, V, W \between l, m, n, p)^2 = 0$, is

$$(GV - HW)(Fxt + Uyz) + (HW - FU)(Gyt + Vzx) + (FU - GV)(Hzt + Wxy) = 0.$$

If the point E be (ξ, η, ζ, τ), it can be shewn that the point F, or $(\xi', \eta', \zeta', \tau')$, is given by

$$\frac{\xi'}{\tau'} = \frac{G\eta - H\zeta}{\eta W - \zeta V}, \quad \frac{\eta'}{\tau'} = \frac{H\zeta - F\xi}{\zeta U - \xi W}, \quad \frac{\zeta'}{\tau'} = \frac{F\xi - G\eta}{\xi V - \eta U}.$$

Further, it can be shewn that these values secure that all the determinants, of three rows and columns, of the array

$$\left\| \begin{matrix} F, & G, & H, & U, & V, & W \\ \eta\zeta, & \zeta\xi, & \xi\eta, & \xi\tau, & \eta\tau, & \zeta\tau \\ \eta'\zeta', & \zeta'\xi', & \xi'\eta', & \xi'\tau', & \eta'\tau', & \zeta'\tau' \end{matrix} \right\|,$$

are zero. It follows thence that the quadric Σ is capable of being written, tangentially, in the form

$$Pl^2 + Qm^2 + Rn^2 + Kp^2 + (l\xi + m\eta + n\zeta + p\tau)^2 + (l\xi' + m\eta' + n\zeta' + p\tau')^2 = 0;$$

this form shews at once that the plane containing any three of the points A, B, C, D, E, F is conjugate, in regard to this quadric, to the plane containing the other three points.

To find the point equation of the quadric represented by this equation, say $\Psi = 0$, we have to find the locus of a point (x, y, z, t), where x is $\partial\Psi/\partial l$, etc., and $lx + my + nz + pt = 0$; that is, writing L for $l\xi + \ldots + p\tau$, and L' for $l\xi' + \ldots + p\tau'$, we have to eliminate l, m, n, p, and ρ, from the equations

$$Pl + \xi L + \xi' L' + \rho x = 0, \text{ etc.,} \quad lx + my + nz + pt = 0.$$

These equations give, if we write

$$U = \frac{x\xi}{P} + \ldots + \frac{t\tau}{K}, \quad U' = \frac{x\xi'}{P} + \ldots + \frac{t\tau'}{K}, \quad S = \frac{x^2}{P} + \ldots + \frac{t^2}{K},$$

$$\omega = 1 + \frac{\xi^2}{P} + \ldots + \frac{\tau^2}{K}, \quad \phi = \frac{\xi\xi'}{P} + \ldots + \frac{\tau\tau'}{K}, \quad \omega' = 1 + \frac{\xi'^2}{P} + \ldots + \frac{\tau'^2}{K},$$

the following equations

$$UL + U'L' + \rho S = 0,$$
$$\omega L + \phi L' + \rho U = 0,$$
$$\phi L + \omega' L' + \rho U' = 0,$$

and hence

$$\begin{vmatrix} U, & U', & S \\ \omega, & \phi, & U \\ \phi, & \omega', & U' \end{vmatrix} = 0,$$

or
$$S(\omega\omega' - \phi^2) - U^2\omega' + 2UU'\phi - U'^2\omega = 0.$$

The expression of the equation of a quadric, referred to a self-polar tetrad, by four squares, has been considered above. The expression of a quadric by five, or more, squares, is considered by Paul Serret, *Géométrie de direction*, Paris, 1869, pp. 265, etc. The theory of the self-conjugate pentads and hexads has been much developed by Reye; see *Crelle's Journal*, Vols. LXXVII (1874), LXXXII (1877), etc.

Ex. 11. Given two arbitrary quadrics, S and Σ, it is possible to find three points of S of which every two are conjugate in regard to Σ, in an infinite number of ways; or, four points of S, forming a tetrad of which every pair of opposite joins consists of conjugate lines in regard to Σ, so that (Ex. 5, above) a fifth point can be found, not necessarily lying on S, such that the five form a self-conjugate pentad, in an infinite number of ways; or, lastly, five points of S, forming five points of a self-conjugate hexad, also in an infinite number of ways. We consider these cases in turn.

In regard to the first case, take an arbitrary point P, on S; then a point Q of S, lying anywhere on the conic in which S is met by the polar plane of P taken in regard to Σ; the polar plane of Q in regard to Σ will pass through P, and will meet the conic spoken of in two points. If R be one of these, then P, Q, R are three points of S of which every two are conjugate to one another, in regard to the quadric Σ. The pole of the plane PQR, say T, which does not necessarily lie on S, will then form, with P, Q, R, a tetrad which is self-polar in regard to Σ.

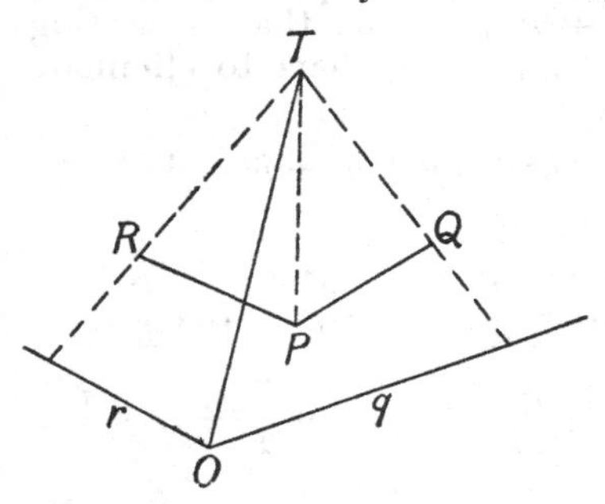

For the self-conjugate pentad, take three arbitrary points P, Q, R lying on S, and the polar lines, in regard to Σ, of the joins QR, RP, PQ; these lines, respectively p, q, r, will then meet in the point, O, which is the pole, in regard to Σ, of the plane PQR. The two planes Qq, Rr will meet in a line, passing through O; and this line will meet the quadric S in two points; let T be one of these. The four points P, Q, R, T, of the quadric S, are then such that the join TQ intersects the polar line, q, in regard to Σ, of the opposite join RP, so that TQ, RP are conjugate lines in regard to Σ. The opposite joins TR, PQ are equally conjugate lines in regard to Σ. Thus (by Ex. 5, above), the remaining pair of opposite joins, TP, RQ, are also conjugate. Wherefore, as has been shewn, the tetrad of points formed by the poles. in regard to Σ, of the planes PQR, TQR, TRP, TPQ, is in perspective with T, P, Q, R, and the centre of perspective, say U, forms, with T, P, Q, R, a self-conjugate pentad in regard to Σ. But U does not necessarily lie upon the quadric S.

For the self-conjugate hexad, take four arbitrary points, P, Q, R, T, of S; then form the quadric $\Phi(P, Q, R, T)$ in regard to Σ, as above explained (Ex. 7); on the curve of intersection, of this quadric Φ, with S, we can then take an arbitrary point E, and from this determine, as above explained, a point F, such that P, Q, R, T, E, F form a self-conjugate hexad, of which the points P, Q, R, T, E, but not necessarily F, are upon S.

Ex. 12. It is very easy now to prove with help of the equation to a quadric, that if upon a quadric S there be, either a self-polar tetrad, or a self-conjugate pentad, or a self-conjugate hexad, in regard to another quadric Σ, then an infinite number, of either of these figures, can be found whose points lie upon S. And though it is desirable, also, to establish this result without help of the symbols, the symbolical proof retains its interest. It depends on the facts we have remarked, that in regard to a self-polar tetrad, or a self-conjugate pentad, or a self-conjugate hexad, the tangential equation of Σ is capable of expression as a sum of squares, respectively four, five, or six in these respective cases.

For brevity let us call a set of h points, (x_r, y_r, z_r, t_r), for $r=1, \ldots, h$, which are so related to the quadric Σ, that the tangential equation of this is capable of a form

$$\sum_{r=1}^{h} A_r (lx_r + my_r + nz_r + pt_r)^2 = 0,$$

involving the squares of the left sides of the equations of these points, a self-conjugate h-ad in regard to Σ. Suppose, then, that a quadric S contains the points of a self-conjugate h-ad in regard to Σ. Suppose further that it has been shewn that $k-1$ points can be taken on S, so related to Σ that they can be completed by another point, not necessarily lying on S, to form a self-conjugate k-ad in regard to Σ; this has been proved, in the preceding example 11, for the cases of $k=4, 5, 6$. It can then be shewn that this remaining point of the k-ad equally lies on S. For if $(\xi_s, \eta_s, \zeta_s, \tau_s)$ be the points of the k-ad, for $s=1, \ldots, k$, we have, by identifying the two forms of which the tangential equation of Σ is capable, an identity, in regard to l, m, n, p, of the form

$$\sum_{r=1}^{h} A_r (lx_r + my_r + nz_r + pt_r)^2 + \sum_{s=1}^{k} P_s (l\xi_s + m\eta_s + n\zeta_s + p\tau_s)^2 = 0;$$

namely, we have the equations of forms such as

$$\sum_r A_r x_r^2 + \sum_s P_s \xi_s^2 = 0, \quad \sum_r A_r y_r z_r + \sum_s P_s \eta_s \zeta_s = 0.$$

If however the point equation of the quadric S be $f(x)=0$, where $f(x) = (a, b, c, d, f, g, h, u, v, w \between x, y, z, t)^2$, these equations lead to

$$\sum_r A_r f(x_r) + \sum_s P_s f(\xi_s) = 0,$$

and we are given that the $h+k-1$ equations, $f(x_r)=0, f(\xi_\sigma)=0$, hold, where $\sigma = 1, \ldots, k-1$. It follows then, supposing that P_k is not zero, that is, supposing that the $k-1$ points of the k-ad which have been taken on S do not form a $(k-1)$-ad, that we also have $f(\xi_k)=0$. So that the quadric S contains the point completing the k-ad, as was to be shewn.

But we see that the condition that there exists upon S a h-ad in regard to Σ may be replaced by another, in order to prove that all the points of the k-ad lie on S. For suppose that the tangential equation of Σ is $(A, B, C, D, F, G, H, U, V, W \between l, m, n, p)^2 = 0$, and, as before, that $k-1$ points of a k-ad in regard to Σ have been taken on S. By identifying the two forms of which the tangential equation of Σ is capable, we have again ten equations of forms such as

$$-A + \sum_s P_s \xi_s^2 = 0, \quad -F + \sum_s P_s \eta_s \zeta_s = 0,$$

and, therefore,
$$-\Theta + \sum_s P_s f(\xi_s) = 0,$$

where

$$\Theta = aA + bB + cC + dD + 2fF + 2gG + 2hH + 2uU + 2vV + 2wW.$$

If then S be so conditioned, relatively to Σ, that $\Theta = 0$, it follows, as before, that all the points of the k-ad lie on S. But, conversely, this same equation shews that, if it be possible to find a k-ad of which all the points lie on the quadric S, then $\Theta = 0$.

When $\Theta = 0$, the quadric S is said to be *outpolar* to Σ.

Ex. 13. If P, P' be two points which are conjugate to one another in regard to a quadric, Σ, and Q, Q' two other such points, and if l, m be the respective polar lines of the lines PP', QQ', in regard to Σ, then a quadric S exists containing the lines l, m and the four points P, P', Q, Q'.

To see this, recall, first, that if planes be drawn through a line, p, the poles of these planes lie on the line, p', which is the polar of p; and the planes through a second line, q, drawn, respectively, conjugate to the former planes, are those which contain these poles; these planes, therefore, form an axial pencil related to that formed by the planes through p.

With the figure we have assumed, the polar plane of the point P' contains both the line l and the point P, namely P' is the pole of the plane lP; thus the planes mP' and lP are conjugate. In this way we see that the planes lP, lP', lQ, lQ' are conjugate, respectively, to the planes mP', mP, mQ', mQ; so that the axial pencil of planes, $l(P, P', Q, Q')$, is related to the axial pencil $m(P', P, Q', Q)$, and, therefore, also to the pencil $m(P, P', Q, Q')$. There is thus (above, p. 4), a quadric surface containing the lines l, m, and the points P, P', Q, Q', of which the transversals, drawn from these four points to meet both l and m, are all generators.

A corollary from this theorem is, that, if P, P', A, A' and Q, Q', B, B' are two tetrads of points, both self-polar in regard to a quadric Σ, then the quadric surface containing the points P, P', A, A', Q, B, B', and the lines AA', BB', passes through the point Q'.

From this can be deduced the theorem proved, with the help of the symbols, in Ex. 12, that every quadric surface containing the seven points P, P', A, A', Q, B, B', contains also the eighth point Q'. But we do not enter into this at present.

Ex. 14. It has been seen (Ex. 7, above) that the lines joining four points A, B, C, D, respectively to the poles of the planes BCD, CAD, ABD, ABC, taken in regard to a quadric with respect to which A, B, C, D have general positions, are generators, of the same system, of another quadric.

Conversely, if four lines, p, q, r, s, of which no two intersect, passing, respectively, through A, B, C, D, be generators of the same system of a quadric, and A', B', C', D' be arbitrary points respectively on p, q, r, s, then there exists a quadric in regard to which the points A', B', C', D' are the poles, respectively, of the planes BCD, CAD, ABD, ABC. In particular, if A', B', C', D' be the points in which these planes are met by p, q, r, s, there exists a quadric touching these planes at these points.

For if the lines p, q, r, s be such generators, a line, s', can be drawn through D to meet all of p, q, r, so that the planes (DA, p), (DB, q) and (DC, r) meet in a line. The condition for this to be so is, that the lines p, q, r should meet the planes BCD, CAD, ABD, respectively, in points whose coordinates, relatively to A, B, C, D, are of the forms $(0, H, G, U)$, $(H, 0, F, V)$, $(G, F, 0, W)$. Considering, similarly, the three lines q, r, s, we see that the line s meets the plane ABC in a point whose coordinates are, with this notation, $(U, V, W, 0)$; the planes (DA, p), (DB, q), (DC, r) meet in a line which meets the plane ABC in the point of coordinates $(F^{-1}, G^{-1}, H^{-1}, 0)$. The quadric whose tangential equation is

$$Fmn + Gnl + Hlm + Ulp + Vmp + Wnp = 0,$$

then touches the planes BCD, etc., in the points where these are met by p, q, r, s; and, if (A, H, G, U), (H, B, F, V), etc., be the coordinates of any points A', B', etc., lying, respectively, on the lines p, q, etc., these points are the poles of the planes BCD, etc., in regard to the quadric whose equation is

$$(A, B, C, D, F, G, H, U, V, W \between l, m, n, p)^2 = 0.$$

We have excluded the consideration of cases in which the lines p, q, etc., lie in the planes ABC, ABD, ACD, BCD.

Dually, if lines l, m, n, k, lying respectively in the planes, BCD, CAD, ABD, ABC, are generators of a quadric, of the same system, and arbitrary planes be drawn through these lines, there is a quadric for which these are the polar planes of A, B, C, D, respectively.

Ex. 15. Consider the twelve points in which the six joins of four points, A, B, C, D, are met by an arbitrary quadric. Three of these, lying on the joins of one of these points, D, to A, B, C, define a plane. We may thus, in thirty-two ways, specify four planes each of which contains three of the points, no two of these planes intersecting in one of the twelve points. If we consider the line in which the plane ABC is met by the plane containing the three points chosen on DA, DB, DC, and the three other lines in which the planes BCD, CAD, ABD are met by the corresponding planes, the four lines so obtained are generators of a quadric, of the same system. (Chasles, *Aperçu historique*, 1837, Note XXXII, p. 400.)

Dually, if we draw the pairs of tangent planes to a quadric from the lines BC, CA, AB, DA, DB, DC, and denote by D' the point of intersection of three planes, one chosen from each of the three pairs of planes drawn from BC, CA, AB, and by A' the point of intersection of three planes drawn from BC, DB, DC, not including the one already taken which is drawn from BC, and so on, so as to obtain four points D', A', etc., each the intersection of three of the twelve tangent planes, then the four lines DD', AA', ... are generators of the same system of a quadric.

For, let A' be the intersection of planes β, γ', δ, drawn, respectively, through DB, DC, BC; let B' be the intersection of planes γ, α', ϵ, drawn, respectively, through DC, DA, CA; and let C' be the intersection of planes α, β', ζ, drawn, respectively, through DA, DB, AB. The planes β, γ', γ, α', α, β', being tangent planes of the quadric, are tangent planes of the quadric cone which is drawn from D to the quadric. Therefore, by Brianchon's theorem (Vol. II, p. 25), the three planes (DA', DA), (DB', DB), (DC', DC) meet in a line, say k; for DA' is the intersection of the planes β, γ', and DA the intersection of the opposite planes α', α; and so on. The lines AA', BB', CC' are, therefore, all met by the line k, which equally meets DD'. These four lines AA', etc., are similarly met by a line through each of the points A, B, C. They are thus such generators as was stated.

Ex. 16. If every intersection of a quadric with a join, such as AD, of four points A, B, C, D, be joined by a plane to the opposite edge, BC, the twelve planes so obtained, touch another quadric. If the point equation of the first quadric be

$$(a, b, c, d, f, g, h, u, v, w \between x, y, z, t)^2 = 0,$$

the tangential equation of the second quadric is

$$\left(\frac{1}{a}, \frac{1}{b}, \frac{1}{c}, \frac{1}{d}, \frac{-f}{bc}, \frac{-g}{ca}, \frac{-h}{ab}, \frac{-u}{ad}, \frac{-v}{bd}, \frac{-w}{cd} \between l, m, n, p\right)^2 = 0.$$

Ex. 17. The transversal is drawn from a point, O, to each pair of opposite joins of four points A, B, C, D, so giving rise to six points. Six other points, one on each join, are similarly obtained from another point, O'. Shew that the twelve points lie on a quadric. If the points O, O' be, respectively, (ξ, η, ζ, τ), $(\xi', \eta', \zeta', \tau')$, referred to A, B, C, D, the quadric has a point equation obtained from

$$\left(\frac{x}{\xi} + \frac{y}{\eta} + \frac{z}{\zeta} + \frac{t}{\tau}\right)\left(\frac{x}{\xi'} + \frac{y}{\eta'} + \frac{z}{\zeta'} + \frac{t}{\tau'}\right) = 0,$$

by changing, in this, the signs of the six terms in $yz, zx, \ldots, yt, zt$. The relation itself is a particular case of that of Ex. 16, when the second quadric there referred to becomes a point-pair; but it will be found to be of importance.

Ex. 18. The point equation of the quadric, touching the planes *BCD*, *CAD*, *ABD*, *ABC*, whose tangential equation is

$$Fmn + \ldots + Ulp = 0,$$

is, by the rule proved above (p. 24),

$$\begin{vmatrix} 0, & H, & G, & U, & x \\ H, & 0, & F, & V, & y \\ G, & F, & 0, & W, & z \\ U, & V, & W, & 0, & t \\ x, & y, & z, & t, & 0 \end{vmatrix}.$$

Prove that, if we put

$$\xi = x\,(FVW)^{\frac{1}{2}},\quad \eta = y\,(GWU)^{\frac{1}{2}},\quad \zeta = z\,(HUV)^{\frac{1}{2}},\quad \tau = t\,(FGH)^{\frac{1}{2}},$$

this is the same as

$$\xi^2 + \eta^2 + \zeta^2 + \tau^2 - 2\,(\eta\zeta + \xi\tau)\cos\alpha - 2\,(\zeta\xi + \eta\tau)\cos\beta \\ - 2\,(\xi\eta + \zeta\tau)\cos\gamma = 0,$$

where $\alpha + \beta + \gamma = \pi$. In fact

$$\cos\alpha = \tfrac{1}{2}\,(GV + HW - FU)\,(GVHW)^{-\frac{1}{2}}, \text{ etc.}$$

Ex. 19. Prove that if (x, y, z, t) be any point of the quadric

$$fyz + gzx + hxy + uxt + vyt + wzt = 0,$$

so also is the point $(Ax^{-1}, By^{-1}, Cz^{-1}, Dt^{-1})$, where $A, = 2fvw$; $B, = 2gwu$; $C, = 2huv$; $D, = 2fgh$, are the minors of the diagonal elements, in the usual determinant of four rows and columns formed with the coefficients of the quadric.

Let P be any point of the quadric, and l, l' be the two generators through P; let the quadric cone be constructed, with vertex at any one, D, of the four points of reference for the coordinates, to contain P, and also the joins, DA, DB, DC, of D to the other three points of reference, and also to contain the generator at D which is of the same system as l; shew that these four cones all meet l again in the same point Q. Making a similar construction for the other system of generators, we obtain a point, Q', on l'. Shew that we can pass from Q to Q' by the above transformation.

Prove that, if the equation of the quadric be put into the form

$$(b - c)\,(axt + yz) + (c - a)\,(byt + zx) + (a - b)\,(czt + xy) = 0,$$

and P be (ξ, η, ζ, τ), one of the lines l, l' is given by

$$x\,(\eta - \zeta) + y\,(\zeta - \xi) + z\,(\xi - \eta) = 0,$$
$$y\,(\zeta - b\tau) - z\,(\eta - c\tau) + t\,(b\eta - c\zeta) = 0,$$

the point Q being given by

$$xt^{-1} = (b\eta - c\zeta)(\eta - \zeta)^{-1}, \quad yt^{-1} = (c\zeta - a\xi)(\zeta - \xi)^{-1},$$
$$zt^{-1} = (a\xi - b\eta)(\xi - \eta)^{-1},$$

and that the other line of l, l' is given by

$$ax(b\eta - c\zeta) + by(c\zeta - a\xi) + cz(a\xi - b\eta) = 0,$$
$$y(\zeta - a\tau) - z(\eta - a\tau) + at(\eta - \zeta) = 0,$$

the point Q' being

$$xt^{-1} = bc(\eta - \zeta)(b\eta - c\zeta)^{-1}, \quad yt^{-1} = ca(\zeta - \xi)(c\zeta - a\xi)^{-1},$$
$$zt^{-1} = ab(\xi - \eta)(a\xi - b\eta)^{-1}.$$

It may also be shewn that, if

$$U = a(b - c)x + b(c - a)y + c(a - b)z,$$
$$V = (a - b)y + (c - a)z + a(b - c)t,$$

the equation of the quadric is capable of the form

$$UV = (by - cz)(y - z)(c - a)(a - b).$$

The coordinates, and equations, of a line, in three dimensions. Elements of the theory of a linear complex. Some elementary results, in regard to lines in space, largely algebraical in character, which it is sometimes convenient to be able to refer to, may be collected here.

Referred to four given points, A, B, C, D, a line may evidently be given in general by the two points in which it meets two, definitely chosen, of the planes BCD, CAD, ABD, ABC. As the position of a point in a plane depends on the two ratios of its three coordinates, we see thus that a line depends upon four parameters, or, in order to avoid the occurrence of infinite parameters, upon the ratios of five parameters. It is desirable to choose these so that the representation shall be valid for all positions of the line; the representation above suggested would fail if the line were entirely contained in one of the two chosen planes. We owe to Cayley (*Coll. Papers*, IV, p. 447, 1860; and VII, p. 66) the discovery that a convenient representation is obtained only by taking the ratios of *six* parameters, which are connected by an equation, reducing them effectively to five.

Let (x_1, y_1, z_1, t_1), (x_2, y_2, z_2, t_2) be the coordinates of any two points of a line. We consider the six quantities

$$l = t_1x_2 - t_2x_1, \quad m = t_1y_2 - t_2y_1, \quad n = t_1z_2 - t_2z_1,$$
$$l' = y_1z_2 - y_2z_1, \quad m' = z_1x_2 - z_2x_1, \quad n' = x_1y_2 - x_2y_1;$$

if, in place of one of the two points, we take any other point of the line, putting, for example, in place of x_2, y_2, z_2, t_2, respectively $\theta x_1 + \phi x_2$, $\theta y_1 + \phi y_2$, etc., it is at once seen that the ratios of l, m, n, l', m', n' remain unaltered. These ratios are therefore given without ambiguity when the line is given. They are evidently connected by the equation

$$ll' + mm' + nn' = 0.$$

Further, it is at once seen that the coordinates, x, y, z, t, of any point of the line, satisfy the four equations $\xi=0$, $\eta=0$, $\zeta=0$, $\tau=0$, where

$$\xi = l't + mz - ny, \quad \eta = m't + nx - lz,$$
$$\zeta = n't + ly - mx, \quad \tau = -(l'x + m'y + n'z),$$

which, when $ll' + mm' + nn' = 0$, are equivalent only to two equations, since we have

$$l\xi + m\eta + n\zeta = 0, \quad l\tau + m'\zeta - n'\eta = 0.$$

The equation $\xi = 0$ is the equation of the plane which joins the line to the reference point A, and similarly for η, ζ, τ, the equation $\tau = 0$ being that of the plane joining the line to the reference point D. As the line can evidently be determined in all cases by two of these four planes, we see, conversely, that the ratios of any six quantities, l, m, n, l', m', n', which are connected by the equation $ll' + mm' + nn' = 0$, suffice to determine a line. It is these quantities which are called the *coordinates of a line* in threefold space.

These coordinates may also be expressed in terms of the coordinates of any two planes which contain the line. If the equations of two such planes be

$$a_1x + b_1y + c_1z + d_1t = 0, \quad a_2x + b_2y + c_2z + d_2t = 0,$$

the line coordinates are given, as regards their ratios, by

$$l = b_1c_2 - b_2c_1, \quad m = c_1a_2 - c_2a_1, \quad n = a_1b_2 - a_2b_1,$$
$$l' = d_1a_2 - d_2a_1, \quad m' = d_1b_2 - d_2b_1, \quad n' = d_1c_2 - d_2c_1.$$

This is seen at once by obtaining, from the equations of the two planes, the equations of the planes joining the line to the four points of reference, A, B, C, D, and comparing these with the forms above expressed in terms of the line coordinates.

It is easily verified that the line coordinates of the six joins of the points of reference are, for DA, $(1, 0, 0;\ 0, 0, 0)$, and, for BC, $(0, 0, 0;\ 1, 0, 0)$, and so on. We may, therefore, symbolically regard any line as represented in terms of these six joins, in the form

$$l(DA) + m(DB) + n(DC) + l'(BC) + m'(CA) + n'(AB).$$

Ex. 1. The conditions that the plane $ax + by + cz + dt = 0$ should contain the line (l, m, n, l', m', n') are two of the following four:

$$ld + m'c - n'b = 0, \quad md + n'a - l'c = 0, \quad nd + l'b - m'a = 0, \quad la + mb + nc = 0.$$

Ex. 2. The plane joining the line (l, m, n, l', m', n') to the point (x, y, z, t) is that whose coordinates are ξ, η, ζ, τ, where, as above, $\xi = l't + mz - ny$, $\tau = -(l'x + m'y + n'z)$, etc. The point of intersection of the line with the plane whose coordinates are (a, b, c, d) is that whose coordinates are

$$ld + m'c - n'b,\ md + n'a - l'c,\ nd + l'b - m'a, \text{ and } -(la + mb + nc).$$

Ex. 3. Expressing the condition that four points, (x_1, y_1, z_1, t_1), (x_2, y_2, z_2, t_2), $(\xi_1, \eta_1, \zeta_1, \tau_1)$ and $(\xi_2, \eta_2, \zeta_2, \tau_2)$, should lie in a plane by the vanishing of a determinant in which the first row consists of x_1, y_1, z_1, t_1, and the third row consists of $(\xi_1, \eta_1, \zeta_1, \tau_1)$, etc., and expanding this determinant as a sum of products of binary determinants formed from the first two rows, and from the third and fourth rows, we see that the condition that the two lines

$$(l_1, m_1, n_1, l_1', m_1', n_1'),\ (l_2, m_2, n_2, l_2', m_2', n_2'),$$

or, say, λ_1 and λ_2, should intersect one another, is $\varpi_{12} = 0$, where

$$\varpi_{12} = l_1 l_2' + l_2 l_1' + m_1 m_2' + m_2 m_1' + n_1 n_2' + n_2 n_1'.$$

When this condition is satisfied, the six quantities such as $\xi l_1 + \eta l_2$, $\xi l_1' + \eta l_2'$, whatever ξ and η may be, satisfy the necessary condition, $\Sigma(\xi l_1 + \eta l_2)(\xi l_1' + \eta l_2') = 0$, that they should be the coordinates of a line. Symbolically, we may say that $\xi\lambda_1 + \eta\lambda_2$ then represents a line. It is easy to see that this is a line passing through the intersection of λ_1 and λ_2, and lying in the plane of these. Similarly, if $\lambda_1, \lambda_2, \lambda_3$ be three lines which meet in a point, not lying in a plane, the line $\xi\lambda_1 + \eta\lambda_2 + \zeta\lambda_3$ is, for proper choice of the ratios $\xi : \eta : \zeta$, any line through this point of intersection.

Ex. 4. If $\lambda_1, \lambda_2, \lambda_3$ be three lines of which no two intersect, the condition that $\xi\lambda_1 + \eta\lambda_2 + \zeta\lambda_3$ should be a line is that ξ, η, ζ should satisfy the condition

$$\varpi_{23}\xi^{-1} + \varpi_{31}\eta^{-1} + \varpi_{12}\zeta^{-1} = 0.$$

There are ∞^1 sets of values of $\xi : \eta : \zeta$ satisfying this condition; they are expressible rationally in terms of a single variable parameter, for instance in the form $\xi = \varpi_{23}$, $\eta = -\varpi_{31}\theta^{-1}$, $\zeta = \varpi_{12}(\theta - 1)^{-1}$. As the condition that two lines should intersect is of the first degree in the coordinates of either line, it follows that any line which intersects $\lambda_1, \lambda_2, \lambda_3$ also intersects $\xi\lambda_1 + \eta\lambda_2 + \zeta\lambda_3$; thus the lines expressed by this last formula are those, of the same system of generators as $\lambda_1, \lambda_2, \lambda_3$, of the quadric determined by these three.

Ex. 5. From Ex. 2, shew that the two tangent planes which can be drawn, from the line (l, m, n, l', m', n'), to touch the quadric whose tangential equation is $f(u, v, w, p) = 0$, where u, v, w, p are the coordinates of a plane, are given by the equation

$$f[l't + mz - ny,\ m't + nx - lz,\ n't + ly - mx,\ -(l'x + m'y + n'z)] = 0.$$

Ex. 6. The polar line of the line (l, m, n, l', m', n'), in regard to the quadric $ax^2 + by^2 + cz^2 + dt^2 = 0$, has the coordinates

$$(bcl', cam', abn', adl, bdm, cdn).$$

Thus the condition that the polar line of the line λ_1, in regard to the quadric, should meet the line λ_2, is

$$\Sigma\,(adl_1l_2 + bcl_1'l_2') = 0,$$

which is symmetrical in regard to the two lines. Two such lines are said to be *conjugate* in regard to the quadric.

Ex. 7. The condition that the line (l, m, n, l', m', n') should touch the quadric $ax^2 + by^2 + cz^2 + dt^2 = 0$ is

$$\Sigma\,(adl^2 + bcl'^2) = 0, \text{ or } \Sigma\,[l\,(ad)^{\frac{1}{2}} + l'\,(bc)^{\frac{1}{2}}]^2 = 0.$$

This is also the condition that the line should intersect its polar line in regard to the quadric.

Ex. 8. If instead of reference points A, B, C, D, we take other points of reference, X, Y, Z, T, the line coordinates, with reference to X, Y, Z, T, of a line whose coordinates relative to A, B, C, D are (l, m, n, l', m', n'), are certain linear functions of l, m, n, l', m', n'. The coefficients in these linear functions are, in fact, the line coordinates, relative to the original points of reference, A, B, C, D, of the six lines which are the joins of the new points of reference, X, Y, Z, T. We may forecast this result if we remark that, by Ex. 4, the condition that a line (l, m, n, l', m', n') should meet the line DA is $l' = 0$, and the condition that the line should meet BC is $l = 0$; if then (L, M, N, L', M', N') be the coordinates of the line referred to X, Y, Z, T, the condition $L = 0$ will be the condition that the line (l, m, n, l', m', n') should meet the line YZ; denoting the coordinates of YZ by $(\lambda_{23}, \ldots, \lambda_{23}', \ldots)$ we may thence expect that L is proportional to $l\lambda_{23}' + l'\lambda_{23} + \ldots$; and similarly for the others.

In fact, in terms of the symbols of X, Y, Z, T, let the symbols of A, B, C, D be, respectively,

$$A = a_1X + a_2Y + a_3Z + a_4T, \ldots, \quad D = d_1X + d_2Y + d_3Z + d_4T,$$

so that, for all values of x, y, z, t,

$$xA + yB + zC + tD = \xi X + \eta Y + \zeta Z + \tau T,$$

where

$$\xi = a_1x + b_1y + c_1z + d_1t, \ldots, \quad \tau = a_4x + b_4y + c_4z + d_4t.$$

Then (ξ, η, ζ, τ) are the coordinates, referred to X, Y, Z, T, of a point whose coordinates referred to A, B, C, D, are (x, y, z, t), and the planes $YZT, \ldots, XYZ$, referred to A, B, C, D, have the respective equations $\xi = 0, \ldots, \tau = 0$. The line coordinates of the line YZ, the intersection of $\xi = 0$, $\tau = 0$, are then given by

$$\lambda_{23} = b_4c_1 - b_1c_4, \ldots, \quad \lambda_{23}' = d_4a_1 - d_1a_4, \ldots.$$

But, if (x_1, y_1, z_1, t_1), (x_2, y_2, z_2, t_2) be two points of the line (l, m, n, l', m', n'), we have

$$L = \tau_1\xi_2 - \tau_2\xi_1 = \begin{vmatrix} a_4x_1 + b_4y_1 + c_4z_1 + d_4t_1, & a_4x_2 + b_4y_2 + c_4z_2 + d_4t_2 \\ a_1x_1 + b_1y_1 + c_1z_1 + d_1t_1, & a_1x_2 + b_1y_2 + c_1z_2 + d_1t_2 \end{vmatrix}$$

$$= \Sigma \begin{vmatrix} x_1, t_1 \\ x_2, t_2 \end{vmatrix} \begin{vmatrix} a_4, d_4 \\ a_1, d_1 \end{vmatrix}$$

$$= \Sigma\,(l\lambda_{23}' + l'\lambda_{23}),$$

which is the formula expected. The value of L' is obtained by similar work in which $(a_2, b_2, \ldots)$, $(a_3, b_3, \ldots)$, respectively, replace $(a_4, b_4, \ldots)$ and $(a_1, b_1, \ldots)$. If then the line coordinates of the line TX be denoted by $\lambda_{41}, = b_2c_3 - b_3c_2$, $\lambda_{41}', = d_2a_3 - d_3a_2$, etc., we find

$$L' = \Sigma\,(l\lambda_{41}' + l'\lambda_{41});$$

and so for the others.

Ex. 9. For the quadric surface whose equation is $x^2 + y^2 - z^2 = t^2$, prove that any generating line of one system has coordinates of the form $(p, q, 1, p, q, -1)$, where $p^2 + q^2 = 1$; and any generating line of the other system has coordinates of the form $(p', q', 1, -p', -q', 1)$, where $p'^2 + q'^2 = 1$. The coordinates of a line of the former system thus satisfy the three conditions $l - l' = 0$, $m - m' = 0$, $n + n' = 0$, whereby, from the ∞^4 existing lines of space there are selected the ∞^1 lines forming the system of generators; for the other system the corresponding conditions are $l + l' = 0$, $m + m' = 0$, $n - n' = 0$.

Ex. 10. *The focal system, and linear complex.* In the course of this chapter we have considered a correspondence between a point and a plane, the polar plane of the point in regard to a quadric, in which to the plane also corresponds the point, this correspondence having the property: (a), that, if the polar plane, ϖ, of any point, P, contain the point Q, then the polar plane of Q also contains P. From this it follows that the pole of any plane through P must lie on ϖ; for if Q' be this pole, since its polar contains P, the polar, ϖ, of P must contain Q'. It also follows that the polars of three points which are in line, must be planes which intersect in a line; for if P' be any point on the line of intersection of the polar planes of two points P and Q, the polar plane of P' must contain both P and Q, and therefore, also, any other point, R, of the line PQ, from which it follows that the polar plane of R contains P'. From this again it follows that the polar points of three planes which meet in a line are three points in line. The correspondence also had the property: (b), that the plane ϖ corresponding to P does not in general contain P, this being so only when the point P is on the quadric of reference. In this relation the coordinates of the plane corresponding to a point, of coordinates (x, y, z, t), are given by formulae of the form

$$ax + hy + gz + ut, \quad hx + by + fz + vt,$$
$$gx + fy + cz + wt, \quad ux + vy + wz + dt,$$

of which the coefficients form a symmetrical system.

We may, however, have a correspondence in which to a point corresponds a plane, and to this plane this point, in which the property (*a*) continues to hold but (*b*) fails, every point lying in the plane which corresponds to it.

Such a correspondence is effected by making correspond to a point (x, y, z, t) the plane whose coordinates are, with any constants a, b, c, a', b', c', the following, respectively,

$$cy - bz - a't, \quad -cx + az - b't, \quad bx - ay - c't, \quad a'x + b'y + c'z,$$

wherein the coefficients form a skew-symmetrical system. The equation of this plane, when X, Y, Z, T are current coordinates, is then

$$X(a't + bz - cy) + Y(b't + cx - az) + Z(c't + ay - bx) - T(a'x + b'y + c'z) = 0,$$

or $$a'(tX - Tx) + b'(tY - Ty) + c'(tZ - Tz) + a(yZ - zY) + b(zX - xZ) + c(xY - yX) = 0,$$

and the correspondence clearly has the properties referred to. It is supposed that, generally, the coefficients $a, b, \ldots$ do not satisfy the condition $aa' + bb' + cc' = 0$. If they do, there is a fixed line of coordinates (a, b, c, a', b', c'), and the correspondence is that by which we make correspond, to any point, the plane joining this point to the fixed line; this is a degenerate relation in that, conversely, the point corresponding to a plane, through the fixed line, is not a definite, but is any, point of this plane.

We may agree to speak of the plane corresponding to any point as the polar plane of that point, and of the point as the pole of the plane. The polar planes of the points of any line are planes which intersect in another line, as we have remarked above; and, dually, the poles of the planes through any line lie on another line. We may agree to speak of either of these lines as the polar line of the other. If P be any point on the polar plane of a point O, the polar plane of P passes through O, and through P; thus the line OP is its own polar line. This is then true of every line, through any point O, lying in the polar plane of O; and is therefore also true of every line, in any plane ϖ, which passes through the pole of ϖ. The line coordinates $(l, m, \ldots)$ of any such self-polar line evidently satisfy the equation

$$a'l + b'm + c'n + al' + bm' + cn' = 0.$$

The aggregate of such self-polar lines is called a *linear complex*; it consists of ∞^3 lines. The aggregate of points with their polar planes, and of lines of the linear complex, may be called a *focal system*.

The representation of the relation takes a very simple form if the points of reference for the coordinates be taken so that two of these points of reference are on the polar line of the line joining the other two points of reference. Suppose these points, A, B, C, D, are such that D and C have for polar planes, respectively, the planes DAB and CAB; the coordinates, of the polar plane of (x, y, z, t), namely

$$a't + bz - cy, \quad b't + cx - az, \quad c't + ay - bx, \quad -(a'x + b'y + c'z),$$

must then reduce, respectively, to $(0, 0, 1, 0)$ and $(0, 0, 0, 1)$, when for (x, y, z, t) we have $(0, 0, 0, 1)$ and $(0, 0, 1, 0)$. From these two facts in turn we deduce $a' = 0 = b'$, and $a = 0 = b$. The plane corresponding to (x, y, z, t) has then the equation

$$c'(tZ - Tz) + c(xY - yX) = 0,$$

and the lines of the linear complex are those for which $c'n + cn' = 0$.

It is clear that the polar plane of any point of a line of the linear complex contains this line. Also that two polar lines do not intersect; for, if they did, the polar plane of any point of one of the lines, as passing through this point and the other line, would contain the first line, which would then be its own polar line and coincide with the other line. Any line which meets two polar lines is evidently a line of the linear complex, since it lies in the polar planes of the two points in which it meets the polar lines.

Now consider any two lines of the linear complex, say λ and λ': the polar plane of any point, P, of λ, is a plane through λ, meeting λ', say, in a point P'. It is clear, or easily verifiable from the equations, that as P describes λ, the range (P'), on λ', described by P', is related to the range (P), on λ, described by (P). Let Q' be the position of P', on λ', corresponding to the position, Q, of P, on λ. Then the polar plane of P contains P', by hypothesis, and contains Q, on the line λ; the polar plane of Q' similarly contains Q and P'. Thus the cross lines PQ' and $P'Q$ are polar lines, one of the other; and therefore, as remarked above, every line meeting PQ' and $P'Q$ is a line of the linear complex. We thus infer that if $A, P, Q, \ldots, D$ and $C, P', Q', \ldots, B$, be related ranges on any two non-intersecting lines in space, the transversal drawn from any point, O, to the lines PQ', $P'Q$, that drawn from O to PR', $P'R$, that drawn from O to QR', $Q'R$, and so on, are in one plane, and are lines of a linear complex in which AB is the polar line of CD. Dually if we join the points in which an arbitrary plane meets PQ' and $P'Q$, all such joins meet in a point of this plane, the pole of the plane in the focal system.

We may thus initiate the theory of the linear complex, as has been remarked in Vol. I, p. 59, without the use of coordinates. Let A, H, D be three points in line, and C, K, B be three points in another line. These are sufficient to determine on the two lines, respectively, two related ranges, $A, H, P, Q, \ldots, D$, and $C, K, P', Q', \ldots, B$; and it is, as there explained, an immediate consequence of the theory of related ranges that the transversals drawn from an arbitrary point, O, to the pairs of cross joins such as (AB, CD), (BH, DK), $(PQ', P'Q)$, lie in a plane through O. This plane is determined by the first two pairs of cross joins, and, therefore, by the points A, H, D and C, K, B alone. In order to establish that the correspondence is that of a focal system, it is necessary to see that if another point, O', be taken in the plane so corresponding to O, then the plane similarly corresponding to O' passes through O, this

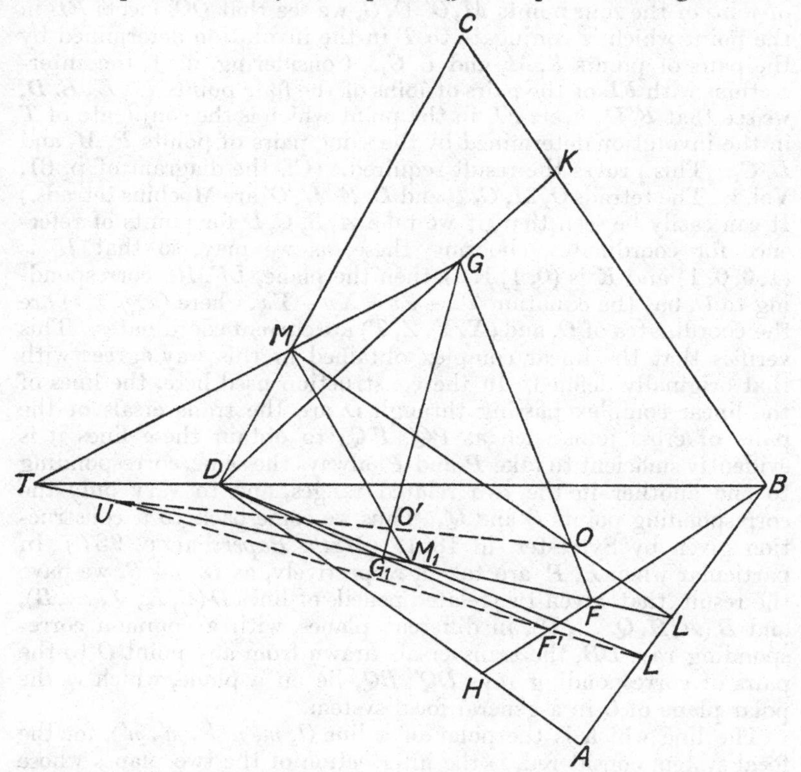

being the condition (a) above remarked. Let the transversal drawn from O to AB and CD meet these, respectively, in L and M; and the transversal drawn from O to BH and DK meet these in F and G respectively. Let the plane of these two transversals meet BD in T. Let O' be any other point of this plane. Let MO' meet FL in M_1, and GO' meet FL in G_1; let DM_1 meet AB in L', and DG_1 meet BH in F'. The point L' is in the plane $O'CD$, and is, therefore, the point in which the transversal from O' to AB, CD meets AB. The point F' is in the plane $O'DK$, and is, therefore, the point in which the transversal from O' to BH, DK meets BH. Thus the plane corresponding to O' is the plane $O'F'L'$, and it is desired to shew that this plane contains O. For this it is necessary and sufficient that the lines OO' and $L'F'$ should meet; their point of meeting will then be on FL, which is the line common to the planes LTM and BAD. What we shew, then, is that OO' and $L'F'$ meet the line FL in the same point. Considering, first, the intersections of FL with the pairs of joins of the four points M, O', O, G, we see that OO' meets FL in the point which is conjugate to T in the involution determined by the pairs of points F, M_1 and L, G_1. Considering, next, the intersections with FL of the pairs of joins of the four points F', L', B, D, we see that $F'L'$ meets FL in the point which is the conjugate of T in the involution determined by the same pairs of points F, M_1 and L, G_1. This proves the result required. (Cf. the diagram of p. 61, Vol. I. The tetrads O, M, G, B and D, F', L', O' are Moebius tetrads.) It can easily be seen that, if we take A, B, C, D for points of reference for coordinates, choosing these, as we may, so that H is $(1, 0, 0, 1)$ and K is $(0, 1, 1, 0)$, then the plane, $LFMG$, corresponding to O, has the equation $Tz - Zt = Xy - Yx$, where (x, y, z, t) are the coordinates of O, and (X, Y, Z, T) are current coordinates. This verifies that the linear complex obtained in this way agrees with that originally defined. In the construction used here, the lines of the linear complex passing through O are the transversals of the pairs of cross joins such as PQ', $P'Q$; to obtain these lines it is evidently sufficient to take P and P' always the same, corresponding to one another in the two related ranges, and to vary only the corresponding points Q and Q'. Thus we come back to a construction given by Sylvester, in 1861. (*Math. Papers*, II, p. 237.) In particular when P, P' are taken, respectively, as D and B, we have the result, that, given two related pencils of lines $D(C, K, Q', \ldots, B)$, and $B(A, H, Q, \ldots, D)$, in different planes, with a common corresponding ray, DB, the transversals drawn from any point O to the pairs of corresponding rays DQ', BQ, lie on a plane, which is the polar plane of O in a general focal system.

The line which is the polar of a line (l, m, n, l', m', n'), for the focal system considered, is the intersection of the two planes whose equations are

$$x(a't_1 + bz_1 - cy_1) + \ldots - t(a'x_1 + b'y_1 + c'z_1) = 0,$$
$$x(a't_2 + bz_2 - cy_2) + \ldots - t(a'x_2 + b'y_2 + c'z_2) = 0,$$

where $l = t_1x_2 - t_2x_1$, $l' = y_1z_2 - y_2z_1$, etc. Its line coordinates are thence easily computed to be

$$\Delta l - Ma, \ldots, \quad \Delta l' - Ma', \ldots$$

where $\quad \Delta = aa' + bb' + cc', \quad M = a'l + \ldots + al' + \ldots.$

The condition for the polar lines to intersect would then be $M^2 = 0$, or the original line would be a line of the linear complex, and coincide with its polar. When this is not so, if we write for the computed line coordinates of the polar line, respectively, $-\Delta\lambda, -\Delta\lambda'$, etc., we may say that $a = M^{-1}(l + \lambda)$, $a' = M^{-1}(l' + \lambda')$, etc.

In particular when the equation of the linear complex is $n' - kn = 0$, the polar line is of coordinates $(l, m, k^{-1}n', l', m', kn)$. Also, the condition, in general, that the polar line of one line, λ_1, should intersect another line, λ_2, is of the form $\Delta\varpi_{12} - M_1 M_2 = 0$, which is symmetrical in regard to the lines.

Consider now two linear complexes. It can be shewn that there are in general two lines which are polars of one another in both the corresponding focal systems. Any transversal of these two polar lines is then a line of both linear complexes. The ∞^2 lines so obtained, which are said to form a *linear congruence*, are then the lines common to the two given linear complexes. If the equations for these two complexes be

$$a'l + \ldots + al' + \ldots = 0, \quad A'l + \ldots + Al' + \ldots = 0,$$

any line common to both satisfies, for all values of σ, the equation

$$(A' + \sigma a')\, l + \ldots + (A + \sigma a)\, l' + \ldots = 0;$$

herein we may choose σ so that the six quantities given by $\lambda = A + \sigma a$, $\lambda' = A' + \sigma a'$, etc., are the coordinates of a line; this requires only

$$AA' + BB' + CC' + \sigma\,(Aa' + A'a + \ldots) + \sigma^2\,(aa' + bb' + cc') = 0,$$

of which there are two different roots in general. This shews that all the lines common to the two complexes meet both of two definite lines, which are then a pair of common polar lines for these.

Incidentally we infer that the ratios of the functions of the coefficients, in the equations of the two linear complexes, which occur in this quadratic equation for σ, are unaltered by change of the reference points for the coordinates, and that these functions must have a geometrical significance. The first and last of these functions are those which vanish when the linear complexes, respectively, reduce to lines all meeting a given line, as we have seen. The condition that the remaining function should vanish,

$$Aa' + A'a + Bb' + B'b + Cc' + C'c = 0,$$

is that either linear complex should consist of pairs of lines which are polars of one another in regard to the other complex; or also, is that, if we take the polar plane of an arbitrary point, in the first focal system, and then the pole of this plane in the other focal system, we reach the same point independently of the order in which the linear complexes are considered. This is easily verified from the general equations. If the complexes be referred to their common pair of polar lines, however, and take the forms $n' - kn = 0$, $n' - Kn = 0$, the polar line $(l, m, K^{-1}n', l', m', Kn)$, of (l, m, n, l', m', n'), in regard to the second focal system, belongs to the first if $Kn = kK^{-1}n'$, which, when $n' = kn$, becomes $K^2 = k^2$, and leads to $K + k = 0$, the complexes not being identical. This however is the form of the condition in question for these particular equations of the complexes. Again, if (ξ, η, ζ, τ) be the pole in $n' - Kn = 0$ of the polar plane of (x, y, z, t) in $n' - kn = 0$, the planes

$$Xy - Yx - k(Tz - Zt) = 0, \quad X\eta - Y\xi - K(T\zeta - Z\tau) = 0,$$

are identical, so that $\xi = x$, $\eta = y$, $\zeta = kK^{-1}z$, $\tau = kK^{-1}t$; for this to be independent of the order in which the complexes are considered, we again require $k^2 = K^2$.

Two linear complexes for which the function $Aa' + A'a + \ldots$ vanishes are said to be *apolar* to one another.

Ex. 11. A focal system is by definition a self dual construct. In particular the pole of the plane $ux + vy + wz + pt = 0$, corresponding to the linear complex $a'l + al' + \ldots = 0$, has the coordinates

$$ap + b'w - c'v, \quad bp + c'u - a'w, \quad cp + a'v - b'u, \quad -(au + bv + cw).$$

Ex. 12. If $P, Q, R, \ldots, P', Q', R', \ldots$, be related ranges on any two skew lines, any line meeting one pair of cross joins, such as PQ' and $P'Q$, meets an infinite number of other pairs of cross joins, such as RS', $R'S$, one of the four points R, S, R', S' being arbitrary.

Ex. 13. Let A, B, C be three points in a line, A', B', C' three points in another line not intersecting the former, and P, Q, R three points in a third line not meeting either of the other two lines. From P let the transversal be drawn to the lines AB', $A'B$, and also the transversal to the lines AC', $A'C$, and the plane of these two transversals be considered. Similarly the plane containing the transversals from Q, to AB', $A'B$, and to AC', $A'C$; and the plane containing the transversals from R, to AB', $A'B$, and to AC', $A'C$. Prove that these three planes meet in a line.

Ex. 14. Take five arbitrary lines in space of which no two intersect. By omitting each in turn we obtain five sets each of four lines; and these four lines have a common pair of transversals. From an arbitrary point a single line can be drawn to meet the two lines of any such pair of transversals. Prove that the five single lines so obtained lie in a plane (Sylvester, *Math. Papers*, II, p. 237).

In fact, the equation of a linear complex being linear in the line coordinates, there is a definite linear complex containing the five given lines. If a line meet two lines of a linear complex, its polar line, in the associated focal system, also meets these two lines; thus the two common transversals of four skew lines of a linear complex are polar lines one of another, and any line meeting both these transversals is also a line of the linear complex.

Ex. 15. If λ be a transversal of three lines, p, q, r, belonging to a linear complex, of which no two intersect, and λ' be the polar line of λ, then λ' is also a transversal of p, q, r, and every line meeting λ and λ' is a line of the linear complex. Conversely, if t be any line of the complex, the two common transversals of p, q, r and t are polar lines of one another. It is easy to prove that the pairs of polar lines which meet p, q, r determine an involution of pairs of points on any one of these lines. Thus a linear complex appears as an aggregate of lines meeting a quadric surface each on two generators, of one system, which are pairs in an involution of these generators. The involution is determined by the three generators p, q, r of the other system taken in turn with two other generators, say t, u, of that system, and the linear complex by the five lines p, q, r, t, u. (Chasles, *Liouville's J.*, IV (1839), p. 348.)

Ex. 16. By expressing the planes which join an arbitrary line to the four points of reference for the coordinates, in terms of two of these planes, and the points in which the line meets the planes containing three of these points of reference in terms of two of these points, shew that the axial pencil of the planes is related to the range of the points (cf. Vol. I, p. 30).

Ex. 17. Prove that the polar plane, in any focal system, of any one, say, D, of the four reference points for coordinates A, B, C, D, contains the poles in this system, of the three planes DBC, DCA, DAB. So that A, B, C, D, and the poles of the four planes such as these, form two Moebius tetrads (cf. Vol. I, pp. 61, 91). Conversely, any two Moebius tetrads define a focal system.

Ex. 18. For the linear complex $al' + a'l +$ etc. $= 0$, the two tetrads of Ex. 17 are both self-polar in regard to the quadric

$$a'bcx^2 + b'cay^2 + c'abz^2 + a'b'c't^2 = 0.$$

Ex. 19. If any two generators of the same system of a quadric be taken, and from each of the points A, B, C, D, forming a self-polar tetrad in regard to the quadric, there be drawn the transversal to these two generators, meeting the planes BCD, ..., ABC respectively in A', ..., D', prove that A, B, C, D and A', B', C', D' are two Moebius tetrads.

Ex. 20. If, in any focal system, A', B', C', D' are the respective poles of the planes BCD, ..., ABC, and U, U', on AA', are harmonic conjugates in regard to A and A', etc., prove that, if U, V, W are in line, the plane $U'V'W'$ contains DD'; but, if the line UVW meets DD', then U', V', W' are also in a line meeting DD', the lines UVW, $U'V'W'$ then meeting DD' in points harmonically conjugate in regard to D and D'. (Cf. p. 144, below.)

Ex. 21. If A, A', B, B', C, C' be any six points, and U, U', on AA', are harmonic conjugates in regard to A and A', while $BVB'V'$, $CWC'W'$ are, similarly, harmonic ranges, prove that, U, V, W being in line, there is no position of this line for which U', V', W' are in line, unless the quadric containing the lines BC, BC', $B'C$, $B'C'$, and the point A, also contains A'.

Ex. 22. Shew that the equations of any four non-intersecting lines can be put into the forms

$$y = px,\ t = qz; \qquad y = -px,\ t = -qz;$$
$$y = p^{-1}x,\ t = q^{-1}z; \quad y = -p^{-1}x,\ t = -q^{-1}z.$$

Taking, then, the four points of respective coordinates $(1, p, m, mq)$, $(1, -p, m, -mq)$, $(p, 1, mq, m)$, $(-p, 1, -mq, m)$, where m is arbitrary, say P, Q, R, S, and also the points P', Q', R', S', whose coordinates are obtained from those of P, Q, R, S, respectively, by changing the sign of m, prove that these form two Moebius tetrads, both self-polar in regard to the quadric $yz - xt = 0$. Further, that the polar plane of any point of one tetrad, in the focal system given by $yz' - y'z + tx' - t'x = 0$, contains three points of the other tetrad, the polar plane of P containing Q', R', S', etc.

Ex. 23. Prove that two of the lines of any given linear complex are generators of the same system of any given quadric surface.

Ex. 24. If we be given a fixed conic, we may take, for any two lines which do not lie in the plane of the conic, the pole, in regard to the conic, of the line joining the points where the lines meet the plane of the conic; and draw from this pole the transversal to the two lines. Now take three lines, p, q, r, not lying in the plane of the conic; let p', q', r' be the transversals, determined as described, respectively of the pairs (q, r), (r, p) and (p, q); then let p'', q'', r'' be the transversals, similarly determined, respectively of the pairs (p, p'), (q, q') and (r, r'). Shew that the transversals so determined, respectively of the pairs (q'', r''), (r'', p''), (p'', q''), coincide in one line. The proof depends on two triads, of points and lines, which are polar reciprocals of one another in regard to the conic (cf. Vol. II, p. 27).

Ex. 25. Shew that, of points of which the coordinates, in terms of a varying parameter θ, are $(\theta^3, \theta^2, \theta, 1)$, there are three lying on an arbitrary plane. These points are therefore said to lie on a cubic curve. Shew that any plane containing the line which is given by the two equations $x - 2y\theta + z\theta^2 = 0$, $y - 2z\theta + t\theta^2 = 0$, meets the curve in two coincident points at the point θ; this line is called the tangent of the curve at this point. Shew that all the tangents of the curve belong to the linear complex expressed by $l + 3l' = 0$.

Ex. 26. Given five points in three dimensions joined in order by five lines, shew that there is a focal system in which each of the points is the pole of the plane containing the two lines which meet in this point, each of the five joins being a line of the associated linear complex.

Ex. 27. Prove that, in regard to a quadric surface expressed in the general form

$$(a, b, c, d, f, g, h, u, v, w \between x, y, z, t)^2 = 0,$$

the line coordinates, $(\lambda, \mu, \nu, \lambda', \mu', \nu')$, of the polar line of the line (l, m, n, l', m', n'), are such that

$$\begin{vmatrix} 0, & -\nu, & \mu, & \lambda' \\ \nu, & 0, & -\lambda, & \mu' \\ -\mu, & \lambda, & 0, & \nu' \\ -\lambda', & -\mu', & -\nu', & 0 \end{vmatrix} = \begin{vmatrix} a, & h, & g, & u \\ h, & b, & f, & v \\ g, & f, & c, & w \\ u, & v, & w, & d \end{vmatrix} \begin{vmatrix} 0, & -n', & m', & l \\ n', & 0, & -l', & m \\ -m', & l', & 0, & n \\ -l, & -m, & -n, & 0 \end{vmatrix} \begin{vmatrix} a, & h, & g, & u \\ h, & b, & f, & v \\ g, & f, & c, & w \\ u, & v, & w, & d \end{vmatrix},$$

where the matrices are to be multiplied by the rule explained in Vol. I, p. 67, the rows of a multiplying matrix being combined with the columns of that multiplied.

Ex. 28. Prove that the tangent lines of the curve of intersection of two quadrics which meet an arbitrary line are eight in number; and that, if the line be a generator of one of the two quadrics, the eight tangents are also generators of this.

The focal system was first investigated by Moebius, in connexion with the reduction of a system of forces acting on a single rigid body, and called by him a *Null* system, as consisting of lines about which the sum of the moments of the forces of the system vanishes (*Lehrbuch der Statik*, 1837). The relations were also studied by Chasles in connexion with the kinematics of a rigid body (1843; *Aperçu historique*, Sec. Edit., 1875, p. 614). On account of these applications the system is often referred to as a *Screw*. The kinematical relations will be considered at a later stage.

CHAPTER II

RELATIONS WITH A FIXED CONIC

SPHERES, CONFOCAL SURFACES; QUADRICS THROUGH THE INTERSECTION OF TWO GENERAL QUADRICS

In the plane geometry of Vol. II we have seen that the ordinary metrical relations are particular cases of the relation of the figure to two arbitrary Absolute points, or, more generally, to an arbitrary Absolute Conic; the recognition of this adds greatly to the breadth of view obtained, without increasing the difficulty of proof. A similar gain is found in the geometry of three dimensions. Here we consider an arbitrary Absolute Conic. In a later Volume we shall consider an arbitrary Absolute Quadric.

When we use coordinates we shall most generally suppose the points of reference, A, B, C, D, so taken that A, B, C form a self-polar triangle in regard to this Absolute Conic, and suppose, therefore, that the equations of this conic have the forms $t = 0$, $x^2 + y^2 + z^2 = 0$. This is immaterial; but it will enable the reader easily to make comparison with metrical formulae which may be familiar.

Parallel lines and planes. Middle point. Lines and planes at right angles. Two lines which meet the plane of the absolute conic in the same point, not themselves lying in this plane, may be said to be *parallel*, relatively to this plane; two planes which meet this absolute plane in the same line may similarly be said to be parallel; a line and a plane may be said to be parallel if they have a common point lying on the absolute plane. Thus, for instance, a plane can be drawn through any line, which shall be parallel to any other given line not meeting the former line.

If the line joining two points A, B, which do not lie in the absolute plane, meet this plane in C, and O be the harmonic conjugate of C in regard to A and B, then O may be spoken of as the *middle point* of A and B, in the relative sense employed.

If two lines, not lying in the absolute plane, meet this plane in two points which are conjugate in regard to the absolute conic, then the lines may be said to be *at right angles*, or *perpendicular* to one another. If a plane and a line meet the absolute plane in a line and a point, respectively, of which the point is the pole of the line in regard to the absolute conic, then the plane and the original line may be said to be at right angles. Similarly, two planes which meet the absolute plane in lines which are conjugate in regard to the absolute conic may be said to be at right angles. Immediate consequences are, that, if two lines are at right angles, a plane can be drawn through either line which shall be at right angles to the other; or, that, if a line and a plane are at right angles, any plane drawn through the line is at right angles to the given plane; or again, that a line can be drawn which shall be at right angles to both of two given lines which do not meet one another; this being the common transversal of the lines, drawn from the point, in the absolute plane, which is the pole of the line joining the points in which the given lines meet the absolute plane.

Ex. 1. Prove that coordinates can be chosen so that the equations of the absolute conic have the forms $t=0$, $x^2+y^2+z^2=0$, and at the same time the equations of any two lines, which do not meet one another, and do not lie in the absolute plane, have the forms

$$lx-my=0,\ z-ct=0, \quad \text{and} \quad lx+my=0,\ z+ct=0.$$

Ex. 2. Shew that the equations of any three lines, of which no two intersect, can be supposed to be of the respective forms

$$(y-bt=0,\ z+ct=0),\ (z-ct=0,\ x+at=0),\ (x-at=0,\ y+bt=0);$$

and that the quadric generated by the transversals of these three lines has the equation

$$ayz+bzx+cxy+abct^2=0.$$

Ex. 3. If between nine quantities (l_1, m_1, n_1), $(l_2, \ldots)$, $(l_3, \ldots)$, there hold the six relations

$$l_1^2+m_1^2+n_1^2=1, \quad l_2^2+m_2^2+n_2^2=1, \quad l_3^2+m_3^2+n_3^2=1,$$
$$l_2l_3+m_2m_3+n_2n_3=0, \quad l_3l_1+m_3m_1+n_3n_1=0, \quad l_1l_2+m_1m_2+n_1n_2=0,$$

then there also hold the six following relations

$$l_1^2+l_2^2+l_3^2=1, \quad m_1^2+m_2^2+m_3^2=1, \quad n_1^2+n_2^2+n_3^2=1,$$
$$m_1n_1+m_2n_2+m_3n_3=0, \quad n_1l_1+n_2l_2+n_3l_3=0, \quad l_1m_1+l_2m_2+l_3m_3=0,$$

while the determinant, of three rows and columns, (l_1, m_2, n_3), has its square equal to 1.

This result, which is often utilised, may be obtained at once by elementary algebra; or, if we use the notation of matrices (explained Vol. I, p. 67), putting Δ for the matrix of three rows whose first row consists of the elements l_1, m_1, n_1, the second and third rows consisting, respectively, of the elements l_2, m_2, n_2 and l_3, m_3, n_3, and $\bar{\Delta}$ for the transposed matrix, obtained from Δ by change of rows into columns, the given equations are equivalent to $\Delta\bar{\Delta}=1$; this leads to $\bar{\Delta}=\Delta^{-1}$, and hence to $\bar{\Delta}\Delta=1$, which is equivalent to the deduced relations. Geometrically, the given relations shew that the conic $x^2+y^2+z^2=0$ is the same as

$$(l_1x+l_2y+l_3z)^2+(m_1x+m_2y+m_3z)^2+(n_1x+n_2y+n_3z)^2=0,$$

so that the lines $l_1x+l_2y+l_3z=0$, etc., determine a self-polar triad for this conic. This proves at once that $l_1m_1+l_2m_2+l_3m_3=0$, etc.

Spheres and Circles. A quadric which meets the absolute plane in the absolute conic is called a sphere, relatively to this conic. A degenerate case is when the quadric consists of the absolute plane and any other plane. Another degenerate case is that in which the sphere is the cone formed by the lines which join an arbitrary point to the points of the absolute conic. The pole of the absolute plane in regard to a sphere is called its *centre*. Let this centre be O, and let any line through O meet the sphere in P and P', and meet the absolute plane in Q. Then O and Q are harmonic conjugates in regard to P and P', or O is the middle point of P, P'. Further, the polar planes, in regard to the sphere, of all points of the line PP', meet in a line, which is, therefore, on the absolute plane, and is the polar line of the point Q in regard to the absolute conic. Thus, the tangent planes of the sphere at P and P' are both at right angles to the line PP'; or, the tangent plane of the sphere, at any point not lying on the absolute conic, is at right angles to the line joining this point to the centre. At a point of the sphere which lies on the absolute conic the tangent plane passes through the pole of the absolute plane, that is, through the centre of the sphere, and all such planes are tangent planes of the tangent cone or cone of contact drawn to the sphere from its centre. (Cf. Chap. I, p. 26.) It may be proved easily that, if P, P' be any two points of a sphere, the plane through the middle point of P, P', at right angles to the line PP', passes through the centre of the sphere.

In the second Volume we have defined a circle, in a plane, as a conic containing two particular points chosen in that plane as absolute points. Let us now agree that for any plane of space the two absolute points shall be the two points in which the plane meets the absolute conic. Then a *circle* will be a conic whose two points of intersection with the absolute plane lie upon the absolute conic. This being understood, consider any plane section of a sphere. Any two plane sections of a quadric are conics with two points in common, the two points where the line of intersection, of the planes of the conics, intersects the quadric. Thus any plane section of a sphere is a circle, relatively to the absolute conic. And, incidentally, it appears, as a quadric can (Chap. I, p. 7) be drawn through two conics which have two points in common, to pass through an arbitrary point, that a sphere can be drawn to contain an arbitrary circle and an arbitrary point. And, hence, that a sphere can be drawn through four arbitrary points in general position.

As in the case of a quadric in general, there pass through any point, P, of a sphere two lines, lying entirely on the sphere, contained in the tangent plane of the sphere at P. These lines meet the absolute plane in two points which, as they lie on the sphere, are on the absolute conic. Any two lines through P, in the tangent plane at P, which are harmonic in regard to these two generating lines of the sphere at P, evidently meet the absolute plane in two points which are conjugate in regard to the absolute conic, and are therefore at right angles to one another. Any plane section of the sphere also meets the two generating lines at P; if this section be projected, from P, on to any plane, ϖ, it becomes, evidently, a conic passing through the two points in which the generating lines of the sphere at P meet the plane ϖ; it becomes then a circle, in the sense above adopted, if the plane ϖ be so taken as to meet the absolute conic in points lying on the tangent plane of the sphere at P. In other words, a plane section of the sphere projects, from any point, P, of the sphere, into a circle, if the plane, upon which the projection is made, is parallel to the tangent plane of the sphere at P.

Ex. 1. Prove that the polar plane of any point in regard to a sphere is at right angles to the line joining that point to the centre of the sphere. This joining line is in fact the polar line of the line in which the absolute plane is met by the polar plane.

Ex. 2. Prove that when the section of the sphere by a plane, ϖ, is projected from a point, H, of the sphere, into a circle on a plane, the centre of this circle is the projection of the pole, in regard to the sphere, of the plane ϖ. The line of intersection of the plane ϖ with the tangent plane of the sphere at H is, in fact, the polar line of the line joining H to the pole of the plane ϖ.

Ex. 3. The line joining any point, Q, of the section of a sphere by a plane, ϖ, to the pole, P, of the plane ϖ, is at right angles to the tangent line of the section at Q. For, a line drawn from P, in the tangent plane of the sphere at Q, meets the sphere in two points lying on the generators of the sphere at Q, and meets the plane ϖ, say in P', upon the tangent plane of the sphere at Q; so that P' is on the tangent line at Q of the section ϖ. But P and P' are harmonic in regard to the points in which the line PP' meets the sphere.

Ex. 4. At either common point of the sections of a sphere by two planes which are conjugate to one another, each plane containing the pole of the other in regard to the sphere, the tangent lines of the sections are at right angles.

Ex. 5. When the two sections of the preceding example are projected, from a point H of the sphere, on to a plane parallel to the tangent plane at H, they become two circles which cut at right angles.

Relations of two spheres. Coaxial spheres. Now consider two spheres. In the first place, they have two points of contact, lying on the absolute conic; that is, there are two points, common to both spheres, on the absolute conic, at each of which the tangent planes of the two spheres are the same. For, let O and O' be the centres of the two spheres, let the line OO' meet the absolute plane in T, and let TA, TB be the tangent lines drawn from T to the absolute conic, touching this in A and B. Then the plane AOT, containing the common tangent line, AT, of the spheres at the common point A, is a tangent plane of both the spheres at this point A; and BOT is, similarly, a common tangent plane at the common point B. This plane equally contains the other centre, O'. The points of contact of the spheres are thus the two points in which either sphere is touched by the two planes which can be drawn to touch it, from the line of centres, OO'. In the next place, the two spheres have in common, beside the absolute conic, a further plane section, which is, therefore, a circle. For, an arbitrary plane cuts the two spheres in circles which, beside their common points lying on the absolute conic, have two further common points; thus the spheres have, beside the absolute conic, an infinite number of points common; let P, Q, R be three of these; then, the plane PQR cuts the spheres in two conics which, beside P, Q, R, have two points in common, those namely where the plane meets the absolute conic; but two conics with five points in common coincide with one another. Thus the two spheres have in common the circle in which the plane PQR meets either; and, therefore, no other common points beside those on the absolute conic. At either of the two points common to this circle and the absolute conic, the tangent plane of either sphere contains the tangent lines of both these conics, and is hence the tangent plane of both spheres. The two points of contact of the spheres are thus the points where their common circular section meets the absolute conic. The plane of this common circular section is called the *radical plane* of the spheres. Through this common circular section can be drawn, as we have seen, an infinite number of further spheres, one through any arbitrary point. These form what is known as a *coaxial system of spheres*. There are, in particular (Chap. I, p. 10), two cones which both contain the absolute conic and the common circle of the two spheres. The vertices of these cones are called the *limiting points* of the system of coaxial spheres. The limiting points lie on the line joining the centres, O and O', of the two original spheres, which also contains the centres of all the spheres of the coaxial system. Further, any sphere drawn through the two limiting points has its centre on the radical plane, and cuts at right angles every sphere of the coaxial system; that is, the tangent plane of such a sphere, at any point common to it and one of the coaxial spheres, is at right angles to the tangent plane of the latter sphere at this point. These facts may be proved from what has been said. We proceed however to utilise the algebraic representation.

Taking reference points A, B, C, D, of which D is arbitrary, but not on the absolute plane, while A, B, C form a self-polar triad in regard to the absolute conic, so that the equations of this conic may be taken to be $t = 0$, $x^2 + y^2 + z^2 = 0$, the equation of a sphere is necessarily of the form

$$a(x^2 + y^2 + z^2) + 2uxt + 2vyt + 2wzt + dt^2 = 0.$$

When the coefficient a vanishes this becomes the aggregate of the absolute plane $t = 0$ and another plane; in general we suppose that a is not zero. This equation is the same as

$$(ax + ut)^2 + (ay + vt)^2 + (az + wt)^2 = (u^2 + v^2 + w^2 - ad)\,t^2,$$

and the point $(u, v, w, -a)$, having for polar the plane $t = 0$, is the centre. The points common to this sphere and any other, of equation

$$a'(x^2 + y^2 + z^2) + 2u'xt + 2v'yt + 2w'zt + d't^2 = 0,$$

satisfy the condition

$$[2(au' - a'u)\,x + 2(av' - a'v)\,y + 2(aw' - a'w)\,z + (ad' - a'd)\,t]\,t = 0;$$

thus, beside the points lying on $t = 0$, there are common points on the radical plane, whose equation is obtained by equating to zero the first factor of the left side. This plane meets $t = 0$ on the line given by $t = 0$ and

$$(au' - a'u)\,x + (av' - a'v)\,y + (aw' - a'w)\,z = 0,$$

which is the polar line, in regard to the absolute conic, of the point $(au' - a'u,\ av' - a'v,\ aw' - a'w,\ 0)$, in which the line joining the centres of the spheres meets the absolute plane; so that the radical plane is at right angles to this joining line. We may thus take the point of intersection of this plane and the line of centres for the reference point D, and the point A to be on the line of centres, so that the radical plane becomes $x = 0$. This choice requires that $v = 0$, $w = 0$, $v' = 0$, $w' = 0$, $ad' - a'd = 0$. Supposing a and a' not to be zero, we may put $a = 1$, $a' = 1$; then $d = d'$ and the equations of the spheres become

$$x^2 + y^2 + z^2 + 2uxt + dt^2 = 0, \quad x^2 + y^2 + z^2 + 2u'xt + dt^2 = 0.$$

Any coaxial sphere, being required to meet $x = 0$ on the circle $x = 0$, $y^2 + z^2 + dt^2 = 0$, will then have an equation

$$x^2 + y^2 + z^2 + 2Uxt + dt^2 = 0;$$

in particular when $U^2 = d$ this represents a cone whose vertex is $(-U, 0, 0, 1)$. The two limiting points are then $(d^{\frac{1}{2}}, 0, 0, 1)$ and $(-d^{\frac{1}{2}}, 0, 0, 1)$. If, for a moment, we put $X = x + d^{\frac{1}{2}}t$, $T = x - d^{\frac{1}{2}}t$, $A = \frac{1}{2}(1 + ud^{-\frac{1}{2}})$, $A' = \frac{1}{2}(1 + u'd^{-\frac{1}{2}})$, $B = 1 - A$, $B' = 1 - A'$, the equations of the two spheres become

$$AX^2 + BT^2 + y^2 + z^2 = 0, \quad A'X^2 + B'T^2 + y^2 + z^2 = 0,$$

which shew that the four points of intersection of the four planes $X = 0$, $T = 0$, $y = 0$, $z = 0$, form a self-polar tetrad for both the spheres, and, therefore, for all the spheres of the coaxial system (Chap. I, p. 33). Herein, however, the planes $y = 0$, $z = 0$ are not unique, being any two planes through the line of centres which are at right angles to one another. The equation of any sphere through the two limiting points is at once seen to be of the form

$$x^2 + y^2 + z^2 + 2Vyt + 2Wzt - dt^2 = 0,$$

having its centre on the radical plane $x = 0$. We proceed to verify that this meets all the spheres of the coaxial system at right angles; and for this purpose we prove that the condition that two spheres, $a(x^2 + y^2 + z^2) + 2uxt + 2vyt + 2wzt + dt^2 = 0$, $a'(x^2 + y^2 + z^2) + \text{etc.} = 0$, should cut at right angles, is

$$2(uu' + vv' + ww') - ad' - a'd = 0.$$

The condition that two planes, of equations $lx + my + nz + pt = 0$, $l'x + \text{etc.} = 0$, should be at right angles, is that the lines

$$lx + my + nz = 0, \quad l'x + \text{etc.} = 0,$$

in which these planes meet $t = 0$, should be conjugate in regard to the absolute conic; that is, $ll' + mm' + nn' = 0$. If now (ξ, η, ζ, τ) be a common point of the two spheres, the respective tangent planes, of such equations as

$$x(a\xi + u\tau) + y(a\eta + v\tau) + z(a\zeta + w\tau) + t(u\xi + v\eta + w\zeta + d\tau) = 0,$$

will be at right angles if

$$(a\xi + u\tau)(a'\xi + u'\tau) + (a\eta + v\tau)(a'\eta + v'\tau) + (a\zeta + w\tau)(a'\zeta + w'\tau) = 0;$$

this, however, is the same as

$$a'S + aS' + [2(uu' + vv' + ww') - ad' - a'd]\,\tau^2 = 0,$$

where $S = 0$, $S' = 0$ are the equations of the two spheres, with (ξ, η, ζ, τ) written for (x, y, z, t). As the point (ξ, η, ζ, τ) satisfies both equations, and is supposed not to lie on the absolute plane, we see that if the spheres cut at right angles at one point, the specified condition is satisfied, and the spheres then cut at right angles at every point; and conversely. The condition is evidently satisfied for any sphere through the limiting points, taken with any one of the coaxial spheres.

Ex. 1. If two circles, not lying in the same plane, have two points in common, a sphere can be described to contain both of them. For these circles together with the absolute conic form three conics of which every two have two points in common; and we have seen (p. 10) that three such conics lie upon a quadric. In the present case this quadric, as containing the absolute conic, is a sphere.

Ex. 2. Prove that the three radical planes, for each pair of three given spheres, meet in a line.

Ex. 3. Given any three circles, no two in the same plane, prove that there is another circle meeting each of them in two points.

If $S_r = 0$ be the equation of a sphere, where

$$S_r = a_r(x^2 + y^2 + z^2) + 2u_r xt + 2v_r yt + 2w_r zt + d_r t^2,$$

it is at once clear that for any six spheres there exists an identity $\lambda_1 S_1 + \lambda_2 S_2 + \ldots + \lambda_6 S_6 = 0$, where $\lambda_1, \ldots, \lambda_6$ are functions of a_r, u_r, v_r, w_r, d_r $(r = 1, \ldots, 6)$. Now let $S_1 = 0$, $S_2 = 0$ be any two chosen spheres through the first of the three given circles, $S_3 = 0$, $S_4 = 0$ be any two chosen spheres through the second of the three given circles, and $S_5 = 0$, $S_6 = 0$ be any two chosen spheres through the third of the three given circles. The equation $\lambda_1 S_1 + \lambda_2 S_2 = 0$ then represents a sphere through the first circle, and the equation $\lambda_3 S_3 + \lambda_4 S_4 = 0$ represents a sphere through the second circle. In virtue of the identity remarked, any point common to these two spheres equally lies on the sphere $\lambda_5 S_5 + \lambda_6 S_6 = 0$, which contains the third circle. Thus the circle common to the first two spheres equally lies on the third. This circle intersects each of the three given circles in two points, since any two circles on the same sphere meet in the two points in which the line of intersection of their planes cuts the sphere.

It may be remarked at once that a similar argument shews that any three lines in space of four dimensions are met by another line. We shall see later how to pass from a circle in three dimensions to a line in four dimensions, or conversely.

Ex. 4. For any two spheres,

$$a(x^2 + y^2 + z^2) + 2uxt + 2vyt + 2wzt + dt^2 = 0,$$
$$a'(x^2 + y^2 + z^2) + 2u'xt + \text{etc.} = 0,$$

let

$$D = u^2 + v^2 + w^2 - ad, \quad E = 2uu' + 2vv' + 2ww' - ad' - a'd,$$
$$D' = u'^2 + v'^2 + w'^2 - a'd';$$

we have shewn that $E = 0$ is the condition for the two spheres to cut at right angles; prove that the first sphere becomes a cone passing through the absolute conic if $D = 0$; also, that the condition that the two spheres should touch, that is, have the same tangent plane, at a point not lying on the absolute plane $t = 0$, is $E^2 = 4DD'$. Further, that the condition that the plane $lx + my + nz + pt = 0$ should touch the first sphere is

$$D(l^2 + m^2 + n^2) - (ul + vm + wn - ap)^2 = 0.$$

Thus, the common tangent planes of the two spheres satisfy the condition

$$D^{-1}(ul + vm + wn - ap)^2 = D'^{-1}(u'l + v'm + w'n - a'p)^2;$$

these common tangent planes are thus the planes drawn to touch either sphere from either of the points whose coordinates are

$$(\lambda u + \lambda' u',\ \lambda v + \lambda' v',\ \lambda w + \lambda' w',\ -\lambda a - \lambda' a'),$$

where λ, λ' are such that $\lambda^2 = 1/D$, $\lambda'^2 = 1/D'$. These two points are called the *centres of similitude* of the two spheres.

Prove further that any sphere, drawn through the vertex of a quadric cone which contains the absolute conic, satisfies the condition $E = 0$, formed for itself and the cone regarded as a degenerate sphere; and may, therefore, be said to cut the cone at right angles.

Shew, too, that, if $l = ua' - u'a$, $l' = vw' - v'w$, etc., the equation of the two common tangent planes of the spheres at the points on the absolute conic at which they touch, is

$$(l't + mz - ny)^2 + (m't + nx - lz)^2 + (n't + ly - mx)^2 = 0.$$

Ex. 5. If two plane sections of a quadric cone be taken, prove that a quadric can be described to contain these, passing through an arbitrary point not lying on either of the planes. Shew also that a sphere can be constructed having a given centre, to pass through a further arbitrary point.

Ex. 6. If O, D be given points, and a variable plane be drawn through D, and a point P taken so that the line OP is at right angles to the plane, and the middle point of O and P lies on the plane, shew that the locus of P is the sphere with centre D which passes through O.

Ex. 7. Four lines in the absolute plane meet in six points, the opposite intersections being conjugate in regard to the absolute conic; any four planes, one through each of these lines, meet, in threes, in the points A, B, C, D. These four points are then such that any two opposite joins of them are two lines at right angles. Prove that, if the line be drawn through each of these points which is at right angles to the plane containing the other three points, these four lines meet in a point, say P. Prove, further, that the six middle points of the joins of the pairs of A, B, C, D lie on a sphere; let its centre be G. Prove that G lies on the line joining P to the centre of the sphere which contains A, B, C, D. A line can be drawn to be at right angles to both of any two opposite joins of

A, B, C, D; taking the intersections of this line with these joins, and the four other points similarly arising for the other two pairs of opposite joins, prove that these six points lie on the sphere containing the middle points of the six joins.

Relations of a quadric in general with the absolute conic. The circular sections. Suppose that the quadric considered does not touch the absolute plane, but meets it in a conic which has four distinct intersections with the absolute conic. These two conics have then a common self-polar triad of points, which we may take for points of reference for coordinates, A, B, C (Vol. II, p. 23). Incidentally also we may remark that if the given quadric be expressed, in regard to any reference points, by an equation wherein the coefficients are real, and the absolute conic, referred to these, be imaginary, then the points A, B, C are real (Vol. II, p. 165). If D be the pole of the absolute plane in regard to the quadric, the four points A, B, C, D will form a self-polar tetrad in regard to the quadric. Referred to A, B, C, D the equation of the quadric will then be of the form

$$\frac{x^2}{a}+\frac{y^2}{b}+\frac{z^2}{c}=t^2,$$

unless the quadric be a cone, consisting of points lying on lines all passing through the point D; in this latter case the equation will be of the form $x^2/a+y^2/b+z^2/c=0$. We may, if we please, include both these equations in the single form $x^2/a+y^2/b+z^2/c=mt^2$.

The point D, the pole of the absolute plane, is called the *centre* of the quadric, or, in the case of the cone, its vertex. The lines DA, DB, DC are called the axes of the quadric; every two of them are at right angles, relatively to the absolute conic. The planes DBC, DCA, DAB are called the *principal planes*. It can now be seen, dealing first with the case when the quadric is not a cone, that through each axis there pass two planes which meet the quadric in sections which are circles, and that all sections in planes parallel to one of these circular sections are also circular sections. For, if P, Q, R, S be the four points common to the quadric and the absolute conic, the section of the quadric by any plane through two of these points is a conic meeting the absolute conic in two points, and is therefore a circle, by the definition. The join of two of these points, P, Q, R, S, contains one of the points A, B, C (Vol. II, p. 23), and a definite plane can be drawn through this join to contain, also, the point D. As two of these joins pass through each of A, B, C, what we have said in regard to circular sections is clear. Of any one of the six sets of parallel planes each meeting the quadric in a circle, there will be two which touch the quadric, these being the tangent planes to the quadric from the line, say, for example, PQ, through which the planes of this set all pass. We thus obtain twelve points of contact, of such tangent planes; these are called *umbilici*. Taking either one of the two generating lines of the quadric at the point P, it is met by one of the two generating lines of the quadric at the point Q, say in U, which will be an umbilicus, while the other generating line at P will be met by the other generating line at Q in another umbilicus, say U'. If B be the one of the three points A, B, C through which PQ passes, the line UU', which is the polar line of PQ, lies in the polar plane of B, that is, in the principal plane DAC, and passes through the centre D, this being the pole of the plane ABC. The twelve umbilici thus lie in fours on the three principal planes, and in pairs on six lines through D. Moreover, as we may consider the intersection of a generating line of the quadric at P with a generating line at Q, or R, or S, we see that the twelve umbilici lie in threes upon the eight generators of the quadric at P, Q, R, S, there being two of these generators through each umbilicus.

If the equation of the quadric be taken in the form remarked above, $\Sigma x^2/a = t^2$, the equations of the absolute conic being $t = 0$, $\Sigma x^2 = 0$, and p, q, r be defined by

$$(a-b)(a-c)p = a, \quad (b-c)(b-a)q = b, \quad (c-a)(c-b)r = c,$$

the point whose coordinates are

$$(a+\lambda)p^{\frac{1}{2}}, \quad (b+\lambda)q^{\frac{1}{2}}, \quad (c+\lambda)r^{\frac{1}{2}}, \quad 1,$$

describes a line, as λ varies. The point of this arising for $\lambda = \infty$, namely $(p^{\frac{1}{2}}, q^{\frac{1}{2}}, r^{\frac{1}{2}}, 0)$, satisfies both the equations $\Sigma x^2/a = 0$, $\Sigma x^2 = 0$, and is thus one of the four points P, Q, R, S; it is at once seen that $\Sigma (a+\lambda)^2/(a-b)(a-c) = 1$, so that the line lies entirely upon the quadric. By taking all the possibilities of sign for $p^{\frac{1}{2}}, q^{\frac{1}{2}}, r^{\frac{1}{2}}$, we thus obtain the eight generators of the quadric which contain the twelve umbilici. In particular, the umbilici are obtained by the values $-a, -b, -c$, of λ.

These eight generators are a set of lines which may be derived from any one of them by a process of harmonic inversion which often occurs, and may be described: if we are given a point O, and a plane ϖ not containing O, we may, from any point P, obtain another point, P', as that point on the line OP which is the harmonic conjugate of P, in regard to O and the point where OP meets the plane ϖ. Then as P describes any line, P' also describes a line, meeting the former on the plane ϖ; and as P describes any plane, P' also describes a plane, meeting the former plane on the plane ϖ. Suppose now we have a tetrad of points, A, B, C, D; we can employ this process of harmonic inversion taking, for O, any one of these four points and, for the associated plane ϖ, the plane containing the other three. Thereby we have four inversions, and we may subject a point to one of these, or to a combination of two or more of them, taken in succession. It is at once seen that thereby only seven new points are obtained; if the original point be of coordinates (x, y, z, t) relative to A, B, C, D, the aggregate of eight points are those of coordinates $(\pm x, \pm y, \pm z, t)$. Similarly from a line, by repeated inversions with this tetrad, eight lines in all are obtained. It is clear from the above expressions that the eight generators containing the umbilici are such a set of eight lines.

Case of a cone. When the quadric reduces to a cone whose vertex is not on the absolute plane, so that its equation is capable of the form $x^2/a + y^2/b + z^2/c = 0$, the equations of the absolute conic being $t = 0$, $x^2 + y^2 + z^2 = 0$, there will still be, in the general case, four intersections P, Q, R, S, of the cone with the absolute conic, and therefore, as before, three principal planes, through the vertex of the cone, the planes $x = 0$, $y = 0$, $z = 0$, intersecting in pairs in the axes of the cone. It is still the case that a plane containing two of the points P, Q, R, S meets the cone in a circle, so that there are six sets of parallel planes each meeting the cone in a circle. The circular sections containing the axes of the cone degenerate however each into two generators of the cone; and the umbilici all coincide with the vertex. A modification of statement is clearly necessary when two or more of the points P, Q, R, S coincide. The case most frequently considered is that in which the cone touches the absolute conic in two distinct points, say, when Q coincides with P and S with R. There are then still triads which are self-polar both in regard to the absolute conic and in regard to the section of the cone by the absolute plane, one point of all of these being the common pole of PR in regard to the two conics, the other two being any pair of points on PR which are harmonic conjugates in regard to P and R. The cone, in this case, has one unique axis, and all sections by planes at right angles to this are circles, so that the cone is said to be a *cone of revolution*, or a *right circular cone*. Any two lines at right angles to one another, through the vertex of the cone, in the plane at right angles to this unique axis, may be regarded as the other two axes. Besides the circular sections spoken of there are no others which are not degenerate. Finally, if the cone has its vertex, C, on the absolute plane, in which case it is called a *cylinder*, there will be two generators of the cone lying in the absolute plane. If these meet the absolute conic in distinct points, respectively, P and R, and Q and S, so that P, C, R are in line, as are Q, C, S, then any plane through either of the four lines PQ, QR, RS, SP, meets the cylinder in a circle. But when P coincides with Q and R coincides with S, or the cylinder touches the absolute plane, no undegenerate circular sections lie in planes through PR. And it is also possible that the vertex of the cone should be on the absolute conic. The necessary modifications may easily be stated. When the cone meets the absolute plane in two distinct lines, its vertex, C, not being on the absolute conic, we may take two points, A and B, on the polar line of C in regard to the absolute conic, which shall be conjugate to one another both in regard to the absolute conic, and in regard to the cone, and take for D a point on the polar line of the absolute plane in regard to the cone; then, with A, B, C, D as point of reference, the equations of the absolute conic may be supposed to be $t=0$, $x^2+y^2+z^2=0$, and the equation of the cone to be of the form $ax^2+by^2=t^2$. This construction fails if the cone touches the absolute plane, its vertex still not being on the absolute conic. If then C be the vertex of the cone, and CA the generator of contact with the absolute plane, the point A being on the polar line of C in regard to the absolute conic, and if B be the pole of CA in regard to the absolute conic, and D be any point not on the absolute plane, the triad A, B, C is self-polar in regard to the absolute conic, whose equations may therefore be supposed to be of the form $t=0$, $x^2+y^2+z^2=0$, while the equation of the cone is of the form $y^2=(Ax+By+Ct)\,t$. This is simplified if we take for CD the other generator of the cone, beside CA, along which the tangent plane contains CB; the equation then reduces to the form $y^2=mxt$. The case when the vertex of the cone lies on the absolute conic may be similarly dealt with.

Ex. Prove that the equation

$$ax^2+by^2+cz^2+2fyz+2gzx+2hxy=0$$

represents a right circular cone (touching the absolute conic in two points), (1) if $f=g=h=0$ and two of a, b, c be equal, (2) if $g=h=0$ and $(a-b)(a-c)=f^2$, (3) when none of f, g, h is zero, if $a-f^{-1}gh=b-g^{-1}hf=c-h^{-1}fg$. Remark also that in general the equation of the cone is capable of the form

$$(a-f^{-1}gh)\,x^2+(b-g^{-1}hf)\,y^2+(c-h^{-1}fg)\,z^2 \\ +fgh\,(f^{-1}x+g^{-1}y+h^{-1}z)^2=0.$$

Quadric touching the absolute plane. When the quadric touches the absolute plane at a point, C, not lying on the absolute conic, there will be two distinct lines of the surface, passing through C, lying in the absolute plane, say these are PCR and QCS, where P, Q, R, S are on the absolute conic. It is then possible to take two points, A and B, on the polar line of C in regard to the absolute conic, which shall be conjugate in regard to this conic, and also harmonic conjugates of one another in regard to the two lines PCR, QCS. From the line AB it is possible to draw another tangent plane to the quadric beside the absolute plane. Let this touch the quadric in the point D. Then, referred to A, B, C, D, the equation of the quadric assumes the form $ax^2 + by^2 + 2wzt = 0$. For, consider the most general form possible

$$ax^2 + by^2 + cz^2 + dt^2 + 2fyz + 2gzx + 2hxy + 2uxt + 2vyt + 2wzt = 0;$$

then, the intersection with the absolute plane is to reduce to two lines, passing through the point C, harmonic in regard to CA, CB, and two such lines are represented by equations of the form $t = 0$, $Ax^2 + By^2 = 0$; we infer, therefore, that $c = f = g = h = 0$; further, the intersection with $z = 0$ is to reduce to two lines given by equations of the form $z = 0$, $lx^2 + 2mxy + ny^2 = 0$; we infer, therefore, that $d = u = v = 0$. This establishes the form in question. The form of the equations of the absolute conic may still be supposed to be $t = 0$, $x^2 + y^2 + z^2 = 0$, the triad A, B, C being self-polar in regard to this. Such a quadric may be called a *paraboloid*. There are four sets, each of parallel planes, meeting the surface in circles, namely those containing any one of the lines PQ, QR, RS, SP. The points of contact of tangent planes, other than the absolute plane, drawn to the quadric through these lines, give four umbilici; these lie in pairs on four generators of the surface passing each through one of the points P, Q, R, S. The case when the quadric touches the absolute plane at a point of the absolute conic may be similarly dealt with.

Ex. 1. For the surface $ax^2 + by^2 + 2wzt = 0$, if $c = a - b$, the umbilici are given by

$$[0, (-ac)^{\frac{1}{2}}, \tfrac{1}{2}c, ab/w] \quad \text{and} \quad [(bc)^{\frac{1}{2}}, 0, \tfrac{1}{2}c, ab/w].$$

Ex. 2. Any two circular sections of a quadric, whose planes are not parallel, lie on a sphere. For two plane sections of a quadric meet in two points, and three conics, of which every two have two points in common, lie on a quadric, as has been remarked.

Ex. 3. The condition for a given plane to be a circular section of a given quadric is that the plane should meet the absolute conic in points lying on the quadric. If the absolute conic be $t = 0$, $x^2 + y^2 + z^2 = 0$, the plane be $lx + my + nz + pt = 0$, and the quadric be in general form $ax^2 + \ldots + 2wzt = 0$, prove that the conditions are the two

$$\frac{bn^2 + cm^2 - 2fmn}{m^2 + n^2} = \frac{cl^2 + an^2 - 2gnl}{n^2 + l^2} = \frac{am^2 + bl^2 - 2hlm}{l^2 + m^2}.$$

Ex. 4. Prove that the axes and asymptotes of a plane section of a quadric are parallel to those of the section by any other parallel plane, and that the centres of these sections, by parallel planes, lie in line. Prove that the axes of the section of the quadric $ax^2 + by^2 + cz^2 = t^2$ by the plane $lx + my + nz = 0$, being the lines through the centre, in this plane, which are conjugate to one another both in regard to the quadric, and in regard to the absolute conic ($t=0$, $x^2+y^2+z^2=0$), are the intersections of the plane with the cone

$$x^{-1}l(b-c)+y^{-1}m(c-a)+z^{-1}n(a-b)=0.$$

Ex. 5. If the general equation $ax^2+\ldots+2wzt=0$ break up into two linear factors, and so represent two planes, the condition that these planes should be at right angles is that the lines in which they meet the absolute plane should form a conic outpolar to the absolute conic (Vol. II, pp. 36, 144). With the usual form for the equations of the absolute conic, the condition for this is $a+b+c=0$.

The pair of tangent planes which can be drawn from the line whose coordinates are (l, m, n, l', m', n'), to the quadric

$$x^2/a+y^2/b+z^2/c=t^2,$$

are at right angles if

$$(b+c)l^2+(c+a)m^2+(a+b)n^2=l'^2+m'^2+n'^2.$$

Ex. 6. Prove that the tangent planes of a quadric at the points of a generator form an axial pencil which is related to the range formed by their points of contact; and that the normals of the surface at the points of the generator, that is the lines through these points at right angles to the tangent planes, meet this generator, and a certain line in the absolute plane, in two related ranges. These normals are therefore generators of a quadric which touches the absolute plane.

If the equation of the given quadric be taken in the form $xU-yV=0$, where U, V are linear in x, y, z, t, the equations of the absolute conic being as before, shew that the normals describe the surface $xV_0+yU_0=0$, where U_0, V_0 are obtained from U, V, respectively, by putting $x=0$ and $y=0$.

The line of striction for one system of generators of a quadric. If two lines which do not intersect one another meet the absolute plane in P and Q, respectively, and R be the pole of the line PQ in regard to the absolute conic, the transversal drawn from R to the given lines is a line at right angles to both, and the only such, as has been remarked. Now suppose the two lines coincide in one generator of a given quadric; the line PQ will then be replaced by the tangent line, at P, of the section of the quadric by the absolute plane. The transversal from R is then replaced by a line from R meeting the quadric in two coincident points; the point of coincidence will therefore be the intersection of the generator with the polar plane of R in regard to the quadric. This construction determines a single point on the generator; the locus of these points, for all the generators of one system, is called the *line of striction* corresponding to the system.

Ex. 1. If the equation of the quadric be $x^2/a^2 + y^2/b^2 - z^2/c^2 = t^2$, the absolute conic being as before, the coordinates of a point of the quadric can be represented in terms of two parameters, λ, μ, in the forms $x/a = 1 - \lambda\mu$, $y/b = \lambda + \mu$, $z/c = 1 + \lambda\mu$, $t = \lambda - \mu$. Then λ is constant for all points of a generator of one system, given by the equations $x/a + z/c = \lambda^{-1}(t + y/b)$, $x/a - z/c = \lambda(t - y/b)$; and μ is constant on the generator of the other system given by

$$x/a + z/c = -\mu^{-1}(t - y/b), \quad x/a - z/c = -\mu(t + y/b).$$

The parameters of the generators of the two systems which meet in a point of the section by $t = 0$ are equal, their point of meeting being of coordinates $a(1 - \lambda^2)$, $2b\lambda$, $c(1 + \lambda^2)$, 0.

Carrying out then the construction above explained it may be shewn that the line of striction, for the generators (λ), meets the generator λ in the point for which μ is given by

$$\mu\lambda(\lambda^2 + A) + A\lambda^2 + 1 = 0, \quad \text{where} \quad A = (2b^{-2} + c^{-2} - a^{-2})/(a^{-2} + c^{-2}).$$

From this it may be seen that the line of striction is a curve meeting an arbitrary plane in four points, and meeting any generator, μ, of the other system, in three points.

Ex. 2. By the same construction, for the quadric represented by the equation $a^{-2}x^2 - b^{-2}y^2 = 2zt$, the tangent line at the point P being however replaced in this case by a generator lying in the absolute plane, shew that the lines of striction are the sections of the surface by the planes $a^{-3}x \pm b^{-3}y = 0$. These sections both touch the absolute plane.

Ex. 3. With the notation above, a generator, λ, of the quadric meeting the absolute plane in P, and R being the pole, in regard to the absolute conic, of the tangent line, at P, of the section of the quadric by the absolute plane, prove that, if RP meet this section again in T, and S be the point of the line of striction which lies on the generator considered, then ST is a generator of the other system.

Ex. 4. Hence shew that if a pair of tangent planes of the quadric, at right angles to one another, be drawn through the given generator, their points of contact being the points H, K of this generator, then the pair H, K belong to an involution of pairs of points on this generator, in which the point S, on the line of striction, is the point conjugate to P.

Ex. 5. Shew that the quartic curve in Ex. 1 lies on a cubic cone whose vertex is $(0, b, 0, 1)$, which intersects the quadric surface also, doubly, in the μ-generator for which $\mu = 0$, namely the line given by $x/a = z/c$, $t = y/b$, this being a double line of the cone.

Shew further that a cubic surface can be drawn through the curve and any two μ-generators of the quadric surface; and that no other quadric surface than the original contains the curve.

In the construction given, as the point P varies, the point R describes a conic in the absolute plane, the polar reciprocal, in regard to the absolute conic, of the conic described by P. The polar planes of the points R envelope a quadric cone, whose vertex is the centre of the quadric. The tangent planes of this cone are thus related, in the sense employed in Vol. I, to the generators of the quadric, through the points P; and the curve is the locus of the intersection of a plane of the cone with the corresponding generator. More generally, if $P=0$, $Q=0$, $R=0$, $S=0$, $U=0$, $V=0$, $W=0$ be arbitrary planes, we may consider the locus of the intersection of the plane $U\theta^2+2V\theta+W=0$, which envelopes the cone $UW=V^2$, with the line $P=\theta R$, $Q=\theta^{-1}S$, which is a corresponding line of the quadric $PQ=RS$. The locus is a curve of this quadric lying on the cubic surface $UP^2+2VPR+WR^2=0$, which also intersects the quadric in the generator $P=0$, $R=0$ counted twice. The curve lies also on the cubic surface

$$U(P-\phi S)(P-\psi S)+V[(P-\phi S)(R-\psi Q)+(P-\psi S)(R-\phi Q)] \\ +W(R-\phi Q)(R-\psi Q)=0$$

which contains two arbitrary generators, ϕ, ψ, of the other system, given by $P=\phi S$, $R=\phi Q$, etc.

The construction for the line of striction is evidently not limited to the case when the generators lie on a quadric surface.

Ex. 6. Notice that the two generators of the surface

$$a^{-2}x^2+b^{-2}y^2-c^{-2}z^2=t^2,$$

at the point (x_0, y_0, z_0, t_0), are the intersections of the tangent plane with the two planes $(xy_0-x_0y)/ab=\pm(zt_0-z_0t)/c$.

The director sphere of a quadric. If three tangent planes of a quadric, which does not touch the absolute plane, be such that every two are at right angles, it can be shewn that the point of intersection of the three planes lies on a definite sphere, whose centre is the centre of the quadric.

We can suppose the equation of any plane put into the form $lx+my+nz=pt$, with l, m, n so chosen that $l^2+m^2+n^2=1$. Further the condition that two planes,

$$l_1x+m_1y+n_1z=p_1t, \quad l_2x+m_2y+n_2z=p_2t,$$

should be at right angles, when the equations of the absolute conic are $t=0$, $x^2+y^2+z^2=0$, is that $l_1l_2+m_1m_2+n_1n_2=0$. If then we have three planes of which every two are at right angles, with equations $l_rx+m_ry+n_rz=p_rt$, we may suppose the six equations $l_r^2+m_r^2+n_r^2=1$, $l_rl_s+m_rm_s+n_rn_s=0$ to be satisfied. Therefore, as remarked above (Ex. 3, p. 71), the six equations such as $l_1^2+l_2^2+l_3^2=1$, $l_1m_1+l_2m_2+l_3m_3=0$, are also satisfied. Wherefore, if (x, y, z, t) be the point of intersection of the planes, from $\Sigma(l_r x + m_r y + n_r z)^2 = t^2 \Sigma p_r^2$, we infer $x^2 + y^2 + z^2 = t^2 \Sigma p_r^2$.

Now suppose the planes all touch the quadric whose equation is $x^2/a + y^2/b + z^2/c = t^2$; the conditions for this are the three of the form $al_r^2 + bm_r^2 + cn_r^2 = p_r^2$; these however lead to $\Sigma p_r^2 = a + b + c$. Thus the point of intersection of the planes lies on the sphere whose equation is $x^2 + y^2 + z^2 = t^2(a + b + c)$.

And it is clear that if the planes touch, not all the same quadric, but, respectively, the three quadrics whose equations are

$$x^2(a+\lambda_1)^{-1} + y^2(b+\lambda_1)^{-1} + z^2(c+\lambda_1)^{-1} = t^2, \text{ etc.,}$$

so that we have $(a+\lambda_r)l_r^2 + (b+\lambda_r)m_r^2 + (c+\lambda_r)n_r^2 = p_r^2$, or $al_r^2 + bm_r^2 + cn_r^2 = p_r^2 - \lambda_r$, then $\Sigma p_r^2 = \Sigma\lambda_r + a + b + c$. In this case the point of intersection of the planes lies on the sphere whose equation is $x^2 + y^2 + z^2 = a + b + c + \Sigma\lambda_r$.

When the quadric touches the absolute plane, and has an equation $ax^2 + by^2 + 2wzt = 0$, the condition that a plane

$$lx + my + nz + pt = 0,$$

should touch the quadric is $a^{-1}l^2 + b^{-1}m^2 + 2w^{-1}np = 0$. If then three planes touch this quadric and intersect in the point (x, y, z, t), we have three equations such as

$$t(a^{-1}l_r^2 + b^{-1}m_r^2) - 2w^{-1}n_r(l_r x + m_r y + n_r z) = 0.$$

If every two of the planes be at right angles, we can thus infer, as above, by adding these equations, that $t(a^{-1} + b^{-1}) - 2zw^{-1} = 0$. The locus of the point of intersection of the three tangent planes thus degenerates into a plane, the absolute plane being, in a certain sense, also part of this. The plane in question passes through the polar line, on the absolute plane, in regard to the absolute conic, of the point of contact of the quadric with the absolute plane.

If we take the general tangential equation of a quadric, for a plane, $lx + my + nz + pt = 0$, touching this, which contains a point (x, y, z, t), the equation

$$t^2(Al^2 + Bm^2 + Cn^2) + D(lx + my + nz)^2 + 2Ft^2mn + \dots$$
$$- 2Utl(lx + my + nz) - \dots = 0$$

holds. If we have three such planes, mutually at right angles, satisfying this equation, we infer, as above, by addition of the three equations, the equations of the absolute conic being as before, that

$$t^2(A + B + C) + D(x^2 + y^2 + z^2) - 2Utx - 2Vty - 2Wtz = 0.$$

This is then the equation of the locus of the point of intersection of the planes; it represents a sphere whose centre is (U, V, W, D), which is the centre of the quadric, the pole of the plane $t = 0$. The locus degenerates into the absolute plane $t = 0$, together with another plane, when $D=0$; this is the condition that the quadric should touch the plane $t=0$. In general, the sphere is called the *director sphere* of the quadric.

We may consider the matter from a less algebraical point of view, if we recall what was said in Chap. I (p. 52) in regard to the condition that one quadric be outpolar to another. The condition that a quadric S', given by a point equation $a'x^2 + \ldots + 2h'xy = 0$, should be outpolar to another quadric Σ given by a tangential equation $Al^2 + \ldots + 2Hlm = 0$, is $a'A + \ldots + 2h'H = 0$. This condition may be proved, as was the corresponding condition for conics (Vol. II, p. 142), to be independent of the choice of the points of reference for coordinates. Evidently if $S_1' = 0$, $S_2' = 0$ be two quadrics both outpolar to Σ, so is any quadric whose equation is $S_1' + \lambda S_2' = 0$; or, if $\Sigma_1 = 0$, $\Sigma_2 = 0$ be two quadrics both inpolar to S', so is any quadric whose equation is $\Sigma_1 + \mu\Sigma_2$. The condition does not apply directly to the case when the inpolar quadric Σ is a cone; there is no unique tangential equation to a cone, just as there is no unique point equation to a conic, regarded as existing in three dimensions. We may however speak of a cone S as being outpolar to a cone Σ of the same vertex as S, when the conic sections of these cones by any plane are in the corresponding relation to one another. The condition for this is easily expressed when the common vertex of the cones is taken as one of the points of reference for coordinates. This being understood, a cone, of vertex P, which is outpolar to the enveloping cone drawn from P to a quadric Σ, is outpolar to Σ; in particular a pair of planes which are conjugate to one another in regard to a quadric Σ, form, together, a degenerate quadric outpolar to Σ. Assuming these results, which may easily be verified, consider again a point, P, from which pass three tangent planes of a quadric, Σ, which are mutually at right angles in regard to the absolute conic, ω. These planes touch the enveloping cone from P to Σ; this cone thus meets the absolute plane in a conic, ϵ, which is inpolar to ω, being touched by three lines in the absolute plane which form a self-polar triad in regard to ω. Thus the conic ω is outpolar to ϵ (Vol. II, p. 33), and the cone $P\omega$, joining P to ω, is outpolar to the enveloping cone $P\epsilon$, and, therefore, by what has been said, is outpolar to the quadric Σ. The director sphere thus appears as the locus of a point P such that the conic $P\omega$, joining P to the absolute conic ω, is outpolar to the quadric. This result is analogous to the fact that the director circle of a conic is the locus of a point which is such that the lines joining it to the two absolute points of the plane of the conic form together a degenerate conic outpolar to the conic, in regard to which they are conjugate lines. If we adopt this definition we may prove that the locus in question is a sphere by remarking that if P and P' be two points, there is a system of coaxial spheres of which P and P' are the limiting points, the cones $P\omega$, $P'\omega$ intersecting in two conics, one, ω, lying on the absolute plane, the other lying in a plane which is at right angles to PP' and passes through the middle point of P, P'. This second plane contains the centre of the quadric; for this statement is the same as that the two planes containing the complete intersection of the two cones, $P\omega$, $P\omega'$, are conjugate to one another in regard to the quadric Σ; now it has been shewn above (p. 75) that if the equations of the two cones be, respectively, $\Gamma=0$, $\Gamma'=0$, the equation of the two planes is of the form $\Gamma-\lambda\Gamma'=0$; as Γ and Γ' are both outpolar to Σ, so then is the aggregate of the two planes. This involves that they are conjugate to one another in regard to the quadric Σ. It follows now easily (cf. Ex. 6, p. 78, above) that P' lies on the sphere constructed to pass through P which has its centre at the centre of the quadric. More generally, if we consider any two spheres which are outpolar to the quadric Σ, these define a system of coaxial spheres. If P and P' be the limiting points of this system, the cones $P\omega$, $P'\omega$, whose equations, as has been seen (p. 75, above) are obtained linearly from the equations of these spheres, are equally outpolar to Σ. Thus P and P' lie on the director sphere. Therefore this sphere cuts both the given spheres at right angles, as we have proved. Thus, all spheres outpolar to Σ cut the director sphere at right angles. This is the analogue in three dimensions of Gaskin's theorem for conics (Vol. II, p. 48).

When a conic ϵ is inpolar to a conic ω, lying in the same plane, if any tangent line of ϵ be taken, and the pole of this in regard to ω, the two tangents of ϵ drawn from this pole are conjugate to one another, as well as to the chosen tangent, in regard to ω. Wherefore, if any tangent plane be drawn to the quadric Σ from a point, P, of the director sphere, and the line at right angles to this be drawn through P, the two tangent planes of the quadric drawn through this line are at right angles to one another. In particular, if we consider a point P common to the director sphere and the quadric Σ, and take for the tangent plane drawn through P that which touches the quadric at P, since a tangent plane of the quadric can be drawn through either generating line of the quadric at P to contain also any other line through P, we infer that the two generating lines of the quadric at P are at right angles to one another. Conversely, it can be shewn that the points of the quadric at which the generators are at right angles lie upon a sphere; this would furnish another definition of the director sphere.

Ex. 1. Let the plane of a conic, ω, meet a quadric, Σ, in the conic σ; let a generator, l, of Σ meet σ in P; let the polar of P in regard to ω meet σ in Q and R. Through each of Q and R there is a generator of Σ which intersects l; let these be m_1 and m_2. The two tangent planes of Σ drawn through QR are those containing, respectively, m_1 and m_2; for, for instance, m_1 is met by the generator through R which is of the same system as l. Any pair of planes through QR which are conjugate to one another in regard to Σ are then harmonic conjugates in regard to the planes (QR, m_1) and (QR, m_2). We can infer that there are two generators m_1, m_2, of the quadric Σ, which intersect a given generator l, and are at right angles to this, and, if these meet l in P_1 and P_2, that the plane at right angles to l which contains the middle point of P_1, P_2, passes through the centre of the quadric. From this it follows that the sphere, whose centre is the centre of the quadric, which contains the point P_1, also contains the point P_2. The generator m_2 is, similarly, met at right angles by another generator beside l, in a point similarly lying on the same sphere; and so on continually.

Ex. 2. When the quadric is that representable, as above (p. 85), in terms of parameters λ, μ, by the equations

$$x/a = 1 - \lambda\mu, \quad y/b = \lambda + \mu, \quad z/c = 1 + \lambda\mu, \quad t = \lambda - \mu,$$

prove that the condition that two generators, (λ), (μ), of different systems, should be at right angles, is

$$a^2(1-\lambda^2)(1-\mu^2) + 4b^2\lambda\mu + c^2(1+\lambda^2)(1+\mu^2) = 0,$$

shewing also that this arises from

$$x^2 + y^2 + z^2 = (a^2 + b^2 - c^2)t^2.$$

Ex. 3. Supposing the equation of the quadric to be

$$x^2/a + y^2/b + z^2/c = t^2,$$

we have shewn above (p. 80) that a generator of the quadric at one of its intersections with the absolute conic is given, varying λ, by the points expressed by

$$\frac{x}{t} = (a+\lambda)p^{\frac{1}{2}}, \quad \frac{y}{t} = (b+\lambda)q^{\frac{1}{2}}, \quad \frac{z}{t} = (c+\lambda)r^{\frac{1}{2}},$$

where $p = a/(a-b)(a-c)$, etc. The point of this line on the absolute plane, $(p^{\frac{1}{2}}, q^{\frac{1}{2}}, r^{\frac{1}{2}}, 1)$, arises by $\lambda = \infty$. This line, meeting the director sphere once at this point, will meet this sphere in one other point. Shew that this is given by $\lambda = 0$, being $(ap^{\frac{1}{2}}, bq^{\frac{1}{2}}, cr^{\frac{1}{2}}, 1)$. Shew further that the cubic equation in θ expressed by

$$a^2p(a+\theta)^{-1} + b^2q(b+\theta)^{-1} + c^2r(c+\theta)^{-1} = 1$$

possesses only the root $\theta = 0$.

Ex. 4. The director sphere of a quadric is the locus of a point such that the cone joining this to the absolute conic is outpolar to the enveloping cone from this to the quadric. More generally, prove that the locus of a point such that the enveloping cone drawn from this to the quadric $x^2+y^2+z^2=mt^2$ is outpolar to the enveloping cone of the quadric $x^2/a+y^2/b+z^2/c=t^2$, is represented by

$$x^2(b+c)+y^2(c+a)+z^2(a+b)+m[x^2+y^2+z^2-(a+b+c)t^2]=0.$$

Dually, prove that a plane whose section with the quadric expressed by $ax^2+by^2+cz^2=t^2$ is outpolar to the section with

$$a_1^{-1}x^2+b_1^{-1}y^2+c_1^{-1}z^2=t^2,$$

touches the quadric expressed by $A^{-1}x^2+B^{-1}y^2+C^{-1}z^2=t^2$,

where $\qquad A=a_1(bb_1+cc_1+1)(aa_1+bb_1+cc_1)^{-1}$, etc.

Ex. 5. If from a point, T, three lines, TP, TQ, TR, can be drawn, to touch a quadric surface E, of which every two are at right angles, the enveloping cone from T to the surface E is outpolar to the cone $T\omega$, joining T to the absolute conic. If the equation of E be $a^{-1}x^2+b^{-1}y^2+c^{-1}z^2=t^2$, the equations of the absolute conic being as before, prove that the plane PQR is a tangent plane of the surface $(a+\lambda)^{-1}x^2+$ etc. $=t^2$, where $\lambda^{-1}+a^{-1}+b^{-1}+c^{-1}=0$, say, of the surface F. The locus of T is then that of the pole, in regard to E, of the tangent plane of F, namely is the polar reciprocal of F in regard to E, whose equation is $a^{-2}(a+\lambda)x^2+$etc. $=t^2$; this locus intersects E on the curve where the common tangent planes of E and F touch E. (Cf. Vol. II, p. 133, and Ex. 2, p. 93, below.)

Confocal quadrics, defined algebraically. The quadrics represented, with different values of λ, by the equation

$$(a+\lambda)^{-1}x^2+(b+\lambda)^{-1}y^2+(c+\lambda)^{-1}z^2=t^2,$$

are said to be confocal. It is perhaps easier to deal with the theory of such surfaces, in the first instance, with the help of the algebraic symbols; a more synthetic treatment is suggested below.

If we replace (x, y, z, t) by the coordinates of a particular point, (x_1, y_1, z_1, t_1), we obtain a cubic equation for the parameter λ; there are thus, in general, three confocals passing through a given point. It is easy to prove that, if a, b, c be real, and (x_1, y_1, z_1, t_1) be real, the three values of λ are real; for definiteness suppose $a>b>c$; it can then be shewn that the three roots $\lambda_1, \lambda_2, \lambda_3$ are such that

$$-a<\lambda_3<-b<\lambda_2<-c<\lambda_1;$$

it is only necessary to assume that, if the cubic polynomial

$$(\lambda+a)(\lambda+b)(\lambda+c)t_1^2-(\lambda+b)(\lambda+c)x_1^2-\ldots-\ldots$$

assumes values of different sign upon the substitution of two real values of λ, then it vanishes for one (or three) intermediate values of λ. Of the three confocal surfaces thus obtained to pass through the point (x_1, y_1, z_1, t_1), every two cut at right angles. For the condition that the tangent planes,

$$(a+\lambda)^{-1}xx_1+(b+\lambda)^{-1}yy_1+(c+\lambda)^{-1}zz_1=tt_1,$$

and $\qquad (a+\mu)^{-1}xx_1+(b+\mu)^{-1}yy_1+(c+\mu)^{-1}zz_1=tt_1,$

should cut at right angles, namely $(a+\lambda)^{-1}(a+\mu)^{-1}x_1^2+\text{etc.}=0$, is obtained at once by subtraction of the two equations

$$(a+\lambda)^{-1}x_1^2+\text{etc.}=t_1^2, \quad (a+\mu)^{-1}x_1^2+\text{etc.}=t_1^2.$$

Thus any two different confocals cut at right angles at every common point.

For every one of the confocal surfaces the tetrad of points, $(1, 0, 0, 0)$, etc., is self-polar. In each of the four planes of this tetrad there is a conic associated with the confocal system; for instance, in the plane $x=0$, the conic $(b-a)^{-1}y^2+(c-a)^{-1}z^2=t^2$, obtainable from the general equation of the confocals by putting $x=0$ and $\lambda=-a$; and so on; the conic in the plane $t=0$ is $x^2+y^2+z^2=0$, obtainable by putting $t=0$ and $\lambda=\infty$. These four conics are called the *focal conics* of the system; each may be regarded as lying on two degenerate surfaces of the confocal system.

When (x_1, y_1, z_1, t_1) are not all real, the three confocals through this point need not be distinct. We have, in particular, remarked above (p. 90) that, when this point is $(ap^{\frac{1}{2}}, bq^{\frac{1}{2}}, cr^{\frac{1}{2}}, 1)$, in which $p=a/(a-b)(a-c)$, etc., the three roots λ of the cubic equation are all $\lambda=0$, the only confocal of the system through this point being $a^{-1}x^2+b^{-1}y^2+c^{-1}z^2=t^2$. More generally, the confocals through any point, P, of coordinates $[(a+\theta)p^{\frac{1}{2}}, (b+\theta)q^{\frac{1}{2}}, (c+\theta)r^{\frac{1}{2}}, 1]$, are given by $\lambda=0$, and $\lambda=\theta$, of which the root θ is a double root, as may easily be verified; for every value of θ this point P lies on a generator, of the surface $a^{-1}x^2+b^{-1}y^2+c^{-1}z^2=t^2$, passing through one of the intersections of this quadric with the conic $t=0$, $x^2+y^2+z^2=0$; this generator, it may easily be seen, touches the surface

$$(a+\theta)^{-1}x^2+\text{etc.}=t^2$$

at the point P. Thus, this generator touches all the confocals (other than that on which it lies); through any point of it there pass only two distinct confocals, the one on which it lies, and that which it touches there; through one point of it, lying on the director sphere of the surface on which it lies, when this director sphere is defined with the focal conic of the plane $t=0$ as absolute conic (cf. p. 90, above), there passes the surface only, of the confocal system. Further, it may easily be verified that through any point, (x_1, y_1, z_1, t_1), where the quadric $a^{-1}x^2+\text{etc.}=t^2$ is met by the cone $a^{-2}x^2+\text{etc.}=0$, there pass, of the confocal system, only this quadric, counting twice, and the surface of parameter $\lambda=(x_1^2+y_1^2+z_1^2)\,t_1^{-2}-(a+b+c)$; at these points the tangent plane of the surface $a^{-1}x^2+\text{etc.}=t^2$ is such as also to touch the conic $t=0$, $x^2+y^2+z^2=0$.

Ex. 1. Prove that the eight generators of the surface

$$a^{-1}x^2+\text{etc.}=t^2$$

at the points where this meets the conic $t=0$, $x^2+y^2+z^2=0$, lie on the surface expressed by the equation

$$[x^2+y^2+z^2-t^2(a+b+c)]^2-4abc\,(a^{-2}x^2+b^{-2}y^2+c^{-2}z^2)\,t^2=0.$$

Ex. 2. Prove that the polar reciprocal of the quadric

$$(a+\lambda)^{-1}x^2+\text{etc.}=t^2,$$

in regard to the quadric $a^{-1}x^2+\text{etc.}=t^2$, that is, the locus of the poles, in regard to the second quadric, of the tangent planes of the first, is the quadric given by

$$a^{-1}x^2+b^{-1}y^2+c^{-1}z^2-t^2+\lambda\,(a^{-2}x^2+b^{-2}y^2+c^{-2}z^2)=0.$$

Consider now the locus of the poles of an arbitrary plane, say of equation $lx+my+nz=pt$, in regard to all the surfaces of the confocal system. It is evidently the line of which the general point has the coordinates $[(a+\theta)\,l, (b+\theta)\,m, (c+\theta)\,n, p]$, for varying values of θ. This line meets the plane $t=0$ in the point $(l, m, n, 0)$, and is at right angles to the given plane, when the absolute conic is the focal conic of the plane $t=0$. The line meets the plane itself in the point for which θ is given by $(a+\theta)\,l^2+\text{etc.}=p^2$; thus there is one surface of the confocal system touching the plane, that namely for which the parameter λ has this particular value of θ. The line may be called the *axis* of the plane.

It may happen that the axis of a plane lies in the plane itself, namely this is so when the condition $(a+\theta)\,l^2+\text{etc.}=p^2$ is satisfied for all values of θ. Then the plane, satisfying

$$l^2+\text{etc.}=0,\quad al^2+\text{etc.}=p^2,$$

touches the focal conic in the plane $t=0$, and touches the surface $a^{-1}x^2+\text{etc.}=t^2$. In this case, however, we also have, for any value of λ, the equation $(a+\lambda)\,l^2+\text{etc.}=p^2$, and the plane touches every surface of the confocal system; it touches, then, also the other three focal conics. The points of contact all lie on the line which is the axis of the plane; this line meets all the focal conics. When the plane has the equation $xp^{\frac{1}{2}}+yq^{\frac{1}{2}}+zr^{\frac{1}{2}}=t$, where, now,

$$p=a/(a-b)\,(a-c),\ \text{etc.},$$

being then the tangent plane of the quadric $a^{-1}x^2+\text{etc.}=t^2$ at the point $(ap^{\frac{1}{2}}, bq^{\frac{1}{2}}, cr^{\frac{1}{2}}, 1)$, considered above, the axis is a generator of this quadric, as we have seen. We can thus make the statements: (*a*) Confocal surfaces have the property that a plane which touches two of them, touches all, the points of contact being in line. Such a plane touches all the focal conics, in points lying on this line. Conversely, a plane touching two of these conics, or touching one of these and one of the surfaces, touches all; (*b*) Through any point four planes of this character can be drawn, these being the common tangent planes of the enveloping cones drawn from this point to two of the surfaces. Exception arises when the point is on one of the focal conics; then two such planes can be drawn through this point, both passing through the tangent line of this conic at this point.

Alternative ways of initiating the system of confocal quadrics. If we take any two quadrics which have a common self-polar tetrad of points, that is, as will be seen, any two quadrics which are not in particular relation, their equations may be supposed to be $X^2+Y^2+Z^2=T^2$ and $A^{-1}X^2+B^{-1}Y^2+C^{-1}Z^2=T^2$. Then, a plane, of equation $lX+mY+nZ=pT$, which touches both these surfaces satisfies the two equations $l^2+m^2+n^2=p^2$, $Al^2+Bm^2+Cn^2=p^2$, and, therefore, satisfies, also, the equation

$$(A+\sigma)\,l^2+(B+\sigma)\,m^2+(C+\sigma)\,n^2=(1+\sigma)\,p^2,$$

whatever σ may be. This plane thus touches the surface

$$P^{-1}X^2+Q^{-1}Y^2+R^{-1}Z^2=T^2,\ \text{where}\ P=(A+\sigma)/(1+\sigma).$$

If we put $x=(A-1)^{-\frac{1}{2}}X$, etc., $t=T$, with $a=(A-1)^{-1}$, etc., and $\lambda=(1+\sigma)^{-1}$, this last equation becomes

$$(a+\lambda)^{-1}x^2+(b+\lambda)^{-1}y^2+(c+\lambda)^{-1}z^2=t^2,$$

which is of the form we have taken to define the system of confocal quadrics.

Or again, suppose we take, in different planes, any two conics, which have no common point; on the line of intersection of the planes of the conics, let A, B be the points which are conjugate to one another in regard to both the conics; let C and D be the poles of this line in regard to the two conics, respectively. Then, referred to A, B, C, D, the equations of the two conics may be supposed to be, respectively,

$$x^2+y^2+z^2=0,\ t=0,\quad\text{and}\quad Ax^2+By^2=t^2,\ z=0;$$

taking c arbitrary we can then take a and b so that $A=(a-c)^{-1}$, $B=(b-c)^{-1}$. Now consider a plane, $lx+my+nz=pt$, which touches both these conics. We then have

$$l^2+m^2+n^2=0,\quad\text{and}\quad (a-c)\,l^2+(b-c)\,m^2=p^2.$$

From these, whatever λ may be, there follows

$$(a+\lambda)\,l^2+(b+\lambda)\,m^2+(c+\lambda)\,n^2=p^2.$$

The plane thus touches any surface whose equation is of the form $(a+\lambda)^{-1}x^2+\text{etc.}=t^2$. Equally, it touches a definite conic in the plane $x=0$, and a definite conic in the plane $y=0$.

Ex. Our original definition of confocal surfaces was for surfaces not touching the absolute plane. If however, as here, we regard as confocal to a given quadric any quadric touching all the tangent planes of this which also touch the absolute conic, we may define quadrics confocal with the surface of equation $a^{-1}x^2 + b^{-1}y^2 = 2zt$, which touches the absolute plane. For the condition that the plane $lx + my + nz = p$ should touch this is $al^2 + bm^2 + 2pn = 0$. The general confocal is then that given tangentially by

$$(a+\kappa)\,l^2 + (b+\kappa)\,m^2 + \kappa n^2 + 2pn = 0\,;$$

it is easily seen that the point equation of this is

$$(a+\kappa)^{-1}x^2 + (b+\kappa)^{-1}y^2 = 2zt + \kappa t^2.$$

The relations of the four focal conics. Begin, as just explained, with two arbitrary conics, having no point in common, lying in different planes; define thereby two other planes, and in each of these a further conic, touched by the common tangent planes of the two given conics; take the four planes as $t=0, z=0, x=0, y=0$, and denote the conics lying therein by $\delta, \gamma, \alpha, \beta$, respectively. It appears then, from what has been said, and is immediately verifiable, that the poles of an arbitrary plane in regard to these conics, that is, the poles, in regard to these conics, of the lines in which the plane intersects the planes of the four conics, respectively, lie in line. Now take the tangent line, say p, at any point, P, of one of the conics α, β, γ, say β; let this line meet the plane of the conic δ in the point T; from T let the tangent lines, TH, TK, be drawn to the conic δ, touching this in H and K. The points P and H are the poles of the plane PTH, respectively, in regard to β and δ; this plane, therefore, touches α and γ at points lying on the line PH. Consider, now, the conics, α', γ', in the plane of the conic δ, which are obtained by projecting the conics α, γ from the point P. Since the plane PHT touches α on PH, the conic α' touches δ at H, as does the conic γ', for the same reason. Similarly, α' and γ' touch δ at K. If we regard δ as the absolute conic, the plane PHK is at right angles to the line PT, that is, to the tangent of the conic β at the point P. The cones joining P to the conics α, γ have thus, both, contact along two generators with the cone joining P to the absolute conic, δ. They are, therefore, what we have above described as right circular cones, or cones of revolution. But, further, as the plane PHT touches all the confocal surfaces derivable from the focal conics, at points lying on the line PH, it similarly follows that the enveloping cone drawn from P to any one of these surfaces is a right circular cone, likewise touching the cone $P\delta$ along the generators PH and PK. Therefore, if we consider the conic in which the plane PHK cuts one of these confocal surfaces, this is touched by the lines PH, PK. If, then, we take, as the two absolute points of any plane, the points in which this plane meets the absolute conic, we have the result that a plane drawn through a point P of one of the focal conics α, β, γ, at right angles to the tangent line of this conic at this point, meets any of the confocal surfaces in a conic of which this point is one focus. When P is at one of the points where the conic, say β, meets the plane of α, (or of γ), the plane PHK, at right angles to the tangent line p, is the plane of this new conic, α. We thus infer that the four points where the plane of one conic, α, of the three α, β, γ, is met by the other two conics, β and γ, are the foci of this conic α.

Ex. 1. Prove that the equation of the enveloping cone drawn from the point, $(\xi, 0, \zeta, 1)$, of the conic β, whose equations are $y = 0$, $(a-b)^{-1}x^2 + (c-b)^{-1}z^2 = t^2$, to the confocal surface of equation $(a+\lambda)^{-1}x^2 + \text{etc.} = t^2$, may be put in the form

$$H[(x-t\xi)^2 + y^2 + (z-t\zeta)^2] - [(a-b)\,x\xi^{-1} + (b-c)\,z\zeta^{-1} - (a-c)\,t]^2 = 0,$$

where $$H = (a+\lambda)(a-b)\,\xi^{-2} - (c+\lambda)(b-c)\,\zeta^{-2}.$$

Ex. 2. Let the line joining the point $(\xi, 0, \zeta, 1)$, or P, of the conic β, to the point $(x, y, 0, 1)$, or Q, of the conic γ, meet the surface $a^{-1}x^2 + b^{-1}y^2 + c^{-1}z^2 = t^2$ in the points L, M. We consider the symbol $(P, Q\,;\, L, M)$, explained Vol. II, p. 166. This is the value of λ_1/λ_2, where λ_1, λ_2 are the roots of the quadratic equation obtained by substituting $(x+\lambda\xi, y, \lambda\zeta, 1+\lambda)$ for the coordinates in the equation of the quadric. It is found that this symbol has a value $F(P)/\Phi(Q)$, where $F(P)$ depends only on P, and $\Phi(Q)$ depends only on Q. In fact, with $m^2 = (a-c)(a-b)^{-1}$,

$$F(P) = (a^{-\frac{1}{2}}\xi + m^{-1})/(a^{-\frac{1}{2}}\xi - m^{-1}), \quad \Phi(Q) = (a^{-\frac{1}{2}}x + m)/(a^{-\frac{1}{2}}x - m).$$

If, then, we take two points Q, say Q_1, Q_2, on the conic γ, the ratio of $(P, Q_1\,;\, L_1, M_1)$ to $(P, Q_2\,;\, L_2, M_2)$ does not vary with the position of P upon the conic β. This result includes that given in Ex. 4, p. 209, of Vol. II.

Verify that if P_1, P_2 be two points of the conic β, and Q_1, Q_2 two points of the conic γ, there are two enveloping cones, of any surface of the confocal system associated with the conics, which touch the four lines P_1Q_1, Q_1P_2, P_2Q_2, Q_2P_1.

Ex. 3. If the absolute points of any plane of space be the intersections of the plane with a single conic, taken as Absolute, prove that through any point there can be drawn six planes whose sections with a given quadric shall have that point as a focus.

If two quadrics touch one another at all the points of a plane section, prove that the tangent plane at an umbilicus of one of these meets the other in a conic of which the umbilicus is one focus.

Ex. 4. The plane at right angles to a focal conic at a point P meets the polar plane of P, in regard to any chosen surface of the confocal system, in a line l. Prove that if any plane through l meet this surface in the conic ζ, the cone joining P to the points of ζ is a right circular cone.

The axes of a confocal system of quadrics. We have seen that the poles of a plane in regard to all the confocal surfaces lie on a line. The lines so determined are of great importance; the necessity, for dynamical purposes, of considering the properties of these lines, is in fact responsible for much of the theory of confocal quadrics. The fundamental property, just stated, depends on the circumstance that the tangential equation of the general confocal surface is of the form $U+\lambda V=0$, where λ is variable; the dual theorem, that the polar planes of a point, in regard to all quadrics whose point equation is of the form $S+\lambda T=0$, pass through a line, will arise below.

By the pole of a plane in regard to one of the focal conics is meant the pole, in regard to this conic, of the line in which the plane meets the plane of the conic. Thus, by what we have seen, the poles of a plane in regard to the surfaces of the confocal system lie on the line joining the poles of this plane in regard to two of the focal conics. This line is then such as to meet the planes of these conics in points, say H and Q, whose polar lines in regard to these conics are two lines which intersect one another (on the line of intersection of the planes of the conics), say in T; and this is a sufficient condition for such a line. The polar lines of T in regard to the conics contain then, respectively, the points H and Q; but, as T is on the line of intersection of the planes of the conics, these polar lines also contain, respectively, the poles, in regard to the conics, of the line of intersection of their planes; let these two poles, which are fixed points, be called B and D. The general construction for a line which is the locus of the poles of a plane in regard to the confocal surfaces is then as follows: Take the pencil of lines through B, in the plane of one of the conics, obtained as the polars, in regard to this conic, of the points, T, of the line of intersection of the planes of the conics; take the pencil of lines through D similarly obtained with the other conic; to any line of the former pencil corresponds a line of the latter, both having T for pole; take any point H of the former line and any point Q of the latter line. The line HQ is such a line as is desired. Such a line depends then on three parameters, one determining the position of T, two others determining the positions of H and Q on definite lines; we say then that there are ∞^3 such lines, the totality of all possible lines being ∞^4, as we have seen in Chap. I. There is in fact one such line arising from any plane.

Ex. A line being given, as in Chap. I, by the equations

$$l'x+m'y+n'z=0,\quad l't+mz-ny=0,$$

etc., the polar lines of the points, $(l, m, n, 0)$, $(0, -n', m', l)$, where the line meets the planes $t=0$, $x=0$, respectively, in regard to the conics $t=0$, $x^2+y^2+z^2=0$ and $x=0$, $(b-a)^{-1}y^2+(c-a)^{-1}z^2=t^2$, intersect on the line $t=0$, $x=0$, provided $all'+bmm'+cnn'=0$. This is then the condition for such lines.

Let such a line be called an *axis of the confocal system*; the plane, of which the line is the locus of poles for the surfaces of the confocal system, if it is not one of the planes touching all the surfaces of the confocal system, touches one of the confocals, and the line passes through the point of contact, and is at right angles to the plane, with respect to any one of the focal conics regarded as absolute conic. The point of contact may be called the *focus* of the plane. If the plane is tangent to all the quadrics of the confocal system, the axis contains all the points of contact, and, in particular, meets each of the focal conics, being a generator of the surface of the confocal system which passes through the point of meeting, as we have seen above. Thus, eight of these axes lie on any quadric of the system, and there are ∞^1 such lines.

The axes of the confocal system which lie in any general plane, ϖ, touch a conic, which touches each of the principal planes $x=0$, $y=0$, $z=0$, $t=0$. This is clear from the construction for such an axis given above. For, let the lines in which the plane ϖ meets the two principal planes there considered, respectively, be the lines m and n; to any point, U, of m there corresponds a line joining it to the point B of the first plane; to this there corresponds a line through D of the related pencil in the second plane, and this meets the line n in a point, V. The ranges (U), (V), respectively on the intersecting lines m and n, being thus related, the line UV, which is one of the axes of the plane ϖ, envelopes a conic touching the lines m and n. On the line of intersection of the two principal planes considered, there are two points, say A and C, such that BA and DA are the polars of C in regard to the conics, respectively, and BAD is a principal plane. If the plane ϖ meet BA and DA respectively in H and Q, the line HQ is tangent to the conic. Similarly the conic touches the principal plane BCD. In fact any line in one of the four principal planes, or through one of the four principal points, is an axis of the confocal system.

The range of four points, in which an axis of the confocal system, lying in the plane ϖ, meets the four principal planes, is thus related to the range in which any other axis in this plane meets these four planes. This range is, however, related to that in which an axis lying in another plane, ϖ', meets the principal planes. For, to any tangent line, t, of the conic in the plane ϖ can be found two tangents of the conic in the plane ϖ' which intersect the former, say t_1' and t_2'; then the lines t and t_1', being axes of the confocal system in the plane (t, t_1'), meet the four principal planes in related ranges. Thus we reach the conclusion that the axes of the confocal system are the lines upon which the intersections with the principal planes are all related ranges, each of four points. The planes joining an axis to the four principal points, A, B, C, D, are then, also, all related axial pencils, of four planes (Vol. I, p. 30); from this it follows that the axes of the confocal system which pass through an arbitrary point lie on a quadric cone containing the points A, B, C, D.

Ex. If the points $(0, -n', m', l)$, $(n', 0, -l', m)$, $(-m', l', 0, n)$, $(l, m, n, 0)$, in which a line whose coordinates are (l, m, n, l', m', n') meets the principal planes, be denoted, respectively, by the symbols P, Q, R, S, we have $lQ = n'S + mP$, $lR = -m'S + nP$. Two such lines are therefore met by the principal planes in related ranges if mm'/nn' be the same for both (Vol. I, p. 154, Vol. II, p. 166). We saw above that the condition for an axis is $all' + bmm' + cnn' = 0$; this, however, in virtue of $ll' + mm' + nn' = 0$, gives $mm'/nn' = (c-a)/(a-b)$.

The lines in space of which the coordinates satisfy a single relation, as here, are said to form a *complex*. We have already briefly considered the linear complex (above, Chap. I). The particular quadratic complex here arising is called the *tetrahedral* complex. It is a general property of quadratic complexes, that the lines of the complex lying in an arbitrary general plane touch a conic, and the lines of the complex passing through an arbitrary general point form a quadric cone. The latter statement appears, if the point be (x_0, y_0, z_0, t_0), by substituting in the quadratic relation connecting l, m, n, l', m', n' which defines the complex, for l, l', etc., respectively, $tx_0 - t_0x$, $yz_0 - y_0z$, etc.; the result, in virtue of $l't_0 = -mz_0 + ny_0$, $m't_0 = -nx_0 + lz_0$, $n't_0 = -ly_0 + mx_0$, is a quadratic relation connecting $tx_0 - t_0x$, $ty_0 - t_0y$, $tz_0 - t_0z$, which is the equation of the cone. The former statement follows in the same way if the line coordinates be defined by the coordinates of two planes, of which, in the proposition considered, the given plane may be taken as one.

For the tetrahedral complex, the cone is given by

$$(b-c)\,x_0l^{-1} + (c-a)\,y_0m^{-1} + (a-b)\,z_0n^{-1} = 0,$$

where $l = tx_0 - t_0x$, etc. The conic, touched by the lines of the complex which lie in the plane $Ax + By + Cz + Dt = 0$, lies on the cone whose equation is

$$[A\,(b-c)\,x]^{\frac{1}{2}} + [B\,(c-a)\,y]^{\frac{1}{2}} + [C\,(a-b)\,z]^{\frac{1}{2}} = 0.$$

It lies also on the cone

$$[(b-c)\,Dt]^{\frac{1}{2}} + [C\,(c-a)\,z]^{\frac{1}{2}} + [B\,(a-b)\,y]^{\frac{1}{2}} = 0.$$

A line is an axis of the system of confocals only if it contains the poles of some plane, taken in regard to two of the surfaces; in this case it contains the poles of this plane in regard to all the confocals. But, if the pole of a plane, in regard to a particular quadric, lie on a line, the polar line of this line, in regard to this quadric, lies in the plane. Thus a line is an axis of the confocals if its polar lines in regard to two of the confocals lie in one plane, in which case its polar lines in regard to all the confocals lie in this plane. This is the same as saying that a line which is an axis of the confocal system is the intersection of the polar planes, in regard to any two of the confocals, of a suitably chosen point. By taking the particular case when one of the two selected confocals is replaced by a focal conic, we reach the conclusion that a line is an axis of the system of confocals if it is at right angles, in regard to one of the focal conics, taken as absolute conic, to its polar line in regard to one of the confocals.

Ex. 1. The polar line of (l, m, n, l', m', n') in regard to the surface $a^{-1}x^2 + \text{etc.} = t^2$ is the line whose coordinates are

$$(al', \ldots, -bcl, \ldots).$$

The condition that these should be at right angles in regard to the absolute conic $t = 0$, $x^2 + y^2 + z^2 = 0$, is $all' + bmm' + cnn' = 0$; and this is also the condition that they should be at right angles in regard to $x = 0$, $(b-a)^{-1}y^2 + (c-a)^{-1}z^2 = t^2$. The condition that the second line should intersect the line

$$[(a+\theta)\, l', \ldots, -(b+\theta)(c+\theta)\, l, \ldots]$$

is also the same.

Ex. 2. If p be the line containing the poles of a plane, ϖ, in regard to the confocals, say the axis of ϖ, meeting ϖ in P, prove that the axes of planes ϵ, drawn through p, lie in the plane ϖ, and meet any two of the principal planes in related ranges, so that they touch a conic touching these planes; further that two of these axes meet in P, and are at right angles to one another in regard to any one of the focal conics, taken as absolute conic; also that any one of these axes is the polar line of p in regard to a properly chosen one of the confocals.

We may consider the locus of the point, in the plane ϖ, which is the intersection of any plane ϵ, drawn through p, with the axis of this plane. If we speak of the conic, in the plane ϖ, which is the envelope of the axes of the confocals lying in this plane, as a parabola, the absolute points of this plane being the intersections of this plane with the focal conic $t = 0$, $x^2 + y^2 + z^2 = 0$, then the point, P, being the intersection of two tangents of the parabola which are at right angles in regard to these absolute points, is on the directrix; and the locus is that of the intersection of a tangent of the parabola with the line drawn at right angles to this tangent from a particular point, P, of the directrix. It may be shewn that this is a curve meeting an arbitrary line of the plane in three points, which passes twice through P.

Ex. 3. If four planes meet in threes in four points, A, B, C, D, which form a self-polar tetrad in regard to a quadric, the lines in which a plane, ϖ, meets these planes intersect in six points, lying on the joins of A, B, C, D, and these six points, it is easily seen, consist of three pairs of points conjugate to one another in regard to the conic section of the quadric by the plane ϖ. This conic is therefore outpolar to any conic in the plane ϖ which touches the four lines spoken of (Vol. II, p. 33).

Thus, in particular, the conic in the plane ϖ, touched by the axes of the confocal quadrics which lie in this plane, is inpolar to the section of any one of the confocals by this plane. Dually, the quadric cone of axes of the confocals which pass through an arbitrary point is outpolar to the enveloping cone drawn from this point to any one of the confocals.

Ex. 4. Shew that if two points, P, Q, of one of the confocals, be such that the lines through these points at right angles respectively to the tangent planes of the quadric at these points, with respect to any one of the focal conics, (called the *normals* of the quadric at these points), meet one another, then PQ is an axis of the confocal system, and conversely.

Ex. 5. For a conic in space, the centre is the pole, in regard to the conic, of the line in which the plane of the conic meets the absolute plane; and the axes of the conic are the lines through its centre at right angles to one another in regard to the absolute conic and also conjugate to one another in regard to the conic itself. Consider now a conic which is the section of one of the confocals by a plane; take, for instance, the principal plane $t=0$ for absolute plane, and the focal conic of this plane for absolute conic; let D be the centre of the quadric, so determined, O the centre of the conic, and OU, OV its axes, the points U, V being on the line, l, in which the plane meets $t=0$. The line DO is then the polar line of l, in regard to the quadric. We desire to see that the axis OU is at right angles to its polar line in regard to the quadric, and, in fact, that the polar line of OU passes through V. For this, as the polar plane of O contains the line UV, or l, it is only necessary to see that the polar plane of U contains V; this is clear because U and V are conjugate in regard to the quadric. Thus a line, which is an axis of the conic section of a quadric of the confocal system by some plane through the line, is an axis of the confocal system. This condition is necessary that a plane may be possible through the line giving a conic section of which the line is an axis.

If we draw through l one of the tangent planes of the quadric, touching it in P, on the line DO, then the polar line of PU is the line PV. These lines are the normals at P of the two confocals, to the given quadric, which pass through P; and are parallel to the axes of any conic section of the given quadric by a plane parallel to the tangent plane at P (that is, intersecting this tangent plane on the absolute plane).

Ex. 6. Taking one of the confocals, an arbitrary plane, and the axis of this plane, the polar line of this axis, in regard to this confocal, lies in the plane, and meets the section of the surface by the plane in two points. Prove that the tangent planes of the surface at these points are at right angles to the plane, and pass through its axis.

Ex. 7. If (l, m, n, l', m', n') satisfy the equation

$$all' + bmm' + cnn' = 0,$$

it is the axis, in regard to the confocal system, of the plane

$$lx + my + nz = pt,$$

where p has the value $mn(b-c)/l'$, or the equal values, $nl(c-a)/m'$, $lm(a-b)/n'$. The normals of the surface $a^{-1}x^2 + \text{etc.} = t^2$, at the points, (x_1), (x_2), where the line meets it, intersect in $(\xi, \eta, \zeta, 1)$, where $a\xi l = px_1x_2$, etc. The plane through the line whose conic section with this surface has the line for an axis has the equation

$$al'(l't + mz - ny) + bm'(m't + nx - lz) + cn'(n't + ly - mx) = 0.$$

Find the equation of the locus of the centres of the conic sections of all the confocals by an arbitrary plane.

Ex. 8. If the normal of a surface, S, of the confocal system, at P, meet the surface S again in P', and meet another confocal, Σ, of the system, in Q and Q', prove that the normals of Σ at Q and Q' intersect in a point lying on the tangent plane of S at P'. Further, that, if Σ vary, the locus of the point of intersection is a conic, in the tangent plane at P'.

Ex. 9. We have seen that the axes of the confocals which lie in a plane touch a conic, and those which pass through a point are generators of a cone. Prove that the points of the conics lying in planes passing through a line (l, m, n, l', m', n') lie on a surface, which is touched by the tangent planes of the cones whose vertices are on this line. Shew that the points of this surface are expressible by two parameters, θ, ϕ, in the forms

$$\beta\gamma x/t = (\theta + a)^2(l'\phi - p)^{-1}, \quad \gamma\alpha y/t = (\theta + b)^2(m'\phi - q)^{-1}, \text{ etc.},$$

where $\alpha = b - c$, $\beta = c - a$, etc., and $p = (mn' - m'n)(l'^2 + m'^2 + n'^2)^{-1}$, etc. See Plücker's *Neue Geometrie des Raumes*, zw. Abth.; Leipzig, 1868, pp. 193 ff.

The normals of a quadric, and of confocal quadrics, which pass through a point. In the case of confocal conics, we considered, starting from an arbitrary point, the aggregate of lines formed by the polars of this point in regard to all the confocals. The lines were tangents of a parabola, touching the two axes of the confocals (Vol. II, p. 120). For the system of confocal surfaces we may similarly consider the aggregate of the polar planes of an arbitrarily taken fixed point, (x_0, y_0, z_0, t_0), in regard to all the surfaces of the system. The equation of the general plane of this aggregate is $(a+\lambda)^{-1}xx_0+(b+\lambda)^{-1}yy_0+(c+\lambda)^{-1}zz_0=tt_0$; particular planes of the aggregate are the principal planes $x=0$, $y=0$, $z=0$, $t=0$, arising respectively for $\lambda=-a$, $\lambda=-b$, $\lambda=-c$, $\lambda=\infty$. Through an arbitrary point there pass three of these planes, which, when the point is (x_0, y_0, z_0, t_0), are mutually conjugate in regard to all the focal conics. Such an aggregate of planes, each determined by a chosen value of a parameter, λ, is called a developable; and, in this case, as three planes of the aggregate pass through an arbitrary point, it is called a cubic developable. If we take a particular quadric of the confocal system, say $a^{-1}x^2+\text{etc.}=t^2$, and then consider the poles, in regard to this confocal, of all the planes of the developable, the points obtained lie upon a curve, of which three points lie in an arbitrary plane, so that the curve is said to be of the third order. It may properly be regarded as the polar reciprocal of the developable in regard to the quadric chosen. The coordinates of the points of this curve are given by

$$[ax_0(a+\lambda)^{-1},\ by_0(b+\lambda)^{-1},\ cz_0(c+\lambda)^{-1},\ t_0],$$

for varying values of λ; it contains the principal points forming the common self-polar tetrad for the confocal surfaces (respectively for $\lambda=-a, -b, -c, \infty$), and also the point (x_0, y_0, z_0, t_0), for $\lambda=0$. The theory of cubic curves in space will be considered below; such a curve can, in an infinite number of ways, be regarded as consisting of all the points common to two quadric surfaces which have also a generator in common, other than the points of this generator, of which only two lie on the curve. The line which joins two points of the curve being called a chord of the curve, the chords which pass through an arbitrary point of the cubic curve in space are the generators of a quadric cone, and the curve may be defined as consisting of the intersection of two of these cones, other than the generator joining their vertices, which is common to both. For the particular cubic curve now under consideration, every chord, being the polar line, in regard to the surface $a^{-1}x^2+\text{etc.}=t^2$, of the intersection of the polar planes of (x_0, y_0, z_0, t_0) in regard to two confocals of the system, is, like this line, an axis of the confocal system. The chords of the curve passing through an arbitrary point of the curve, (x_1, y_1, z_1, t_1), are then the axes of the confocal system through this point, lying on the cone whose equation is

$$(b-c)\,x_1(xt_1-x_1t)^{-1}+(c-a)\,y_1(yt_1-y_1t)^{-1}+\text{etc.}=0.$$

Consider now a point common to the surface, $a^{-1}x^2 + \text{etc.} = t^2$, and the cubic curve. The tangent plane of the surface at this point will be the polar plane of (x_0, y_0, z_0, t_0) in regard to one of the confocals; thus (x_0, y_0, z_0, t_0) will lie on the axis of this plane. This axis is at right angles to the plane, with respect to every one of the focal conics taken as absolute conic, and is the normal of the surface $a^{-1}x^2 + \text{etc.} = t^2$ at the point of contact of the plane. The normals of this surface which can be drawn through the point (x_0, y_0, z_0, t_0) are then the lines joining this point to the intersections of this surface with the cubic curve. The number of such intersections, of a cubic curve with a quadric surface, is in fact six. In the present case they may be determined from the roots of the sextic equation

$$\frac{ax_0^2}{(a+\lambda)^2} + \frac{by_0^2}{(b+\lambda)^2} + \frac{cz_0^2}{(c+\lambda)^2} = t_0^2.$$

Ex. 1. If a line be drawn from the point (x_0, y_0, z_0, t_0), at right angles, in regard to the focal conic $t = 0$, $x^2 + y^2 + z^2 = 0$, to any plane, ϖ, drawn through the centre of the surface $a^{-1}x^2 + \text{etc.} = t^2$, such that this line meets the diameter of this plane (namely the polar line of the line in which the plane ϖ meets the plane $t = 0$), prove that the locus of the point of meeting is the cubic curve just described.

Ex. 2. Shew that the two lines of Ex. 1, that through (x_0, y_0, z_0, t_0), and the diameter, are such as to meet any plane in two related plane systems (in the sense of Vol. I, p. 148). In general, if two corresponding points, P_1 and P_2, of two related plane systems, which may be in different planes, be joined, respectively, to two fixed points O_1, O_2, of arbitrary position, shew that, for positions of P_1 for which O_1P_1 and O_2P_2 meet, the locus of the point of meeting is a cubic curve.

Ex. 3. Shew that the plane containing the three points (λ_1), (λ_2), (λ_3), of the cubic curve just described, has the equation

$$\frac{x}{ax_0}\frac{(a+\lambda_1)(a+\lambda_2)(a+\lambda_3)}{(a-b)(a-c)} + \frac{y}{by_0}\frac{(b+\lambda_1)(b+\lambda_2)(b+\lambda_3)}{(b-c)(b-a)} + \frac{z}{cz_0}\frac{(c+\lambda_1)(c+\lambda_2)(c+\lambda_3)}{(c-a)(c-b)} = \frac{t}{t_0}.$$

Ex. 4. Shew that the six roots of the sextic equation which gives the six points of the quadric $a^{-1}x^2 + \text{etc.} = t^2$ at which the normals pass through (x_0, y_0, z_0, t_0), are such that

$$\prod_{r=1}^{6}(a+\lambda_r) = -a(a-b)^2(a-c)^2 x_0^2 t_0^{-2},$$

with similar values for the products of the quantities $b + \lambda_r$, and of the quantities $c + \lambda_r$; hence, denoting these six points by

$N_1, N_2, \ldots, N_6$, prove that, if the equation of the plane $N_1N_2N_3$ be written in the form $a^{-\frac{1}{2}}lx + \text{etc.} = t$, the equation of the plane $N_4N_5N_6$ will be of the form $a^{-\frac{1}{2}}l^{-1}x + \text{etc.} = -t$.

Ex. 5. Through every point of the cubic curve there is a line, called the tangent of the curve, such that any plane through this line meets the curve in two points coincident at this point, and in only one other point. Shew that the tangent at the point (x_0, y_0, z_0, t_0) is the line drawn through this point at right angles to the polar plane of this point in regard to the surface $a^{-1}x^2 + \text{etc.} = t^2$. In general only one cubic curve can be drawn to contain six given points. The present cubic curve is thus determined by the facts that it passes through the principal points, the common self-polar tetrad for the confocals, passes through (x_0, y_0, z_0, t_0), and has the line described, as the tangent line at (x_0, y_0, z_0, t_0).

Ex. 6. The chords $N_1N_2, N_1N_3, \ldots, N_1N_6$ of the cubic curve, being axes of the confocal system, lie upon the quadric cone, of vertex N_1, containing the axes of the confocals which pass through N_1. This cone is independent of the point, O, of the normal at N_1, from which the other normals ON_2, etc., are drawn. Thus, as O varies on the normal at N_1, the points N_2, etc., vary on the curve of intersection of the surface $a^{-1}x^2 + \text{etc.} = t^2$ with the cone in question.

Ex. 7. Taking an arbitrary plane section of the surface

$$a^{-1}x^2 + \text{etc.} = t^2,$$

and a fixed point N_1 of this section, there are two other points of this section which are such that the normals of the surface at these points meet the normal of the surface at N_1. For, if the normal at N_2 meet the normal at N_1, the line N_1N_2 is an axis of the confocals; and the axes which lie in this plane touch a conic lying in this plane; two of them, therefore, pass through N_1, say N_1N_2 and N_1N_3; the normals of the surface at the points N_2 and N_3 of the section intersect the normal at N_1, generally in points which do not coincide. If, however, these points coincide, the normals at N_2N_3 intersect one another, and the line N_2N_3 is an axis of the confocals. When this is so, the conic, Σ, of the plane, enveloped by all the axes of the confocals which lie in this plane, is triangularly inscribed in the conic section of the quadric by the plane. This conic, Σ, we have seen, is always inpolar to the other. When both facts are true, the two conics are in the particular relation discussed in Vol. II, p. 149, each being both inpolar and outpolar to the other. It may be proved that, if the equation of the plane be $Ax + By + Cz = Dt$, the condition for this particular relation is

$$\Sigma\,(b-c)^2\,(bcB^2C^2 - aA^2D^2) = 0\,;$$

and that, then, the locus of the intersection of the existing triads of concurrent normals is a line.

The normals of the confocals which are at right angles to a given line, and the polars of this line in regard to the confocals. Consider a line l, which we suppose not to be an axis of the confocal system; in particular, it does not lie in one of the four principal planes nor pass through one of the four principal points. These possibilities are considered below. Every plane, ϖ, drawn through l determines a further line, λ, the axis of the plane ϖ; every confocal, S', also determines a further line, l', the polar line of l in regard to S'. The pole, in regard to S', of any plane through l, in particular of the plane ϖ, lies on l'; the poles of the plane ϖ, in regard to all the surfaces of the confocal system, in particular in regard to the surface S', lie on the axis λ. Thus the lines l' and λ have a point in common. If we take different surfaces S', the corresponding lines l' do not intersect, or l would be one of the axes of the confocals. By considering in the first instance three surfaces S', and the corresponding lines l', we see that the axes, λ, of all the planes, ϖ, drawn through l, are the generators, of one system, of a quadric surface; and, then, that the polar lines l', for all the confocal surfaces, are the generators of the other system of this surface. The system of polar lines l' will have a line in each of the principal planes, the polars, namely, in regard to the focal conics, respectively, of the point in which the line l meets these planes. The quadric surface in question thus touches each of the principal planes.

Considering any particular two of the polar lines l', the ranges in which these are met by all the lines λ will be related to one another, and these both related to the axial pencil of planes ϖ from which the lines λ are determined. For, let X, T be the intersections of the line l with two of the principal planes, respectively; and x, t the polar lines of these points, respectively, with regard to the focal conics in these planes. Any plane, ϖ, through l, will meet these planes, respectively, in two lines, say p and s, passing through X and T, and the poles, say P and S, of these lines, in regard to the appropriate conics, will be points of the lines x and t, respectively. As ϖ varies, the lines p, s will describe related pencils in their respective planes, and the points P, S will, therefore, describe related ranges. The line λ, corresponding to a plane ϖ, is the join PS; this line, therefore, meets x and t in ranges related to the axial pencil of planes ϖ; and it is a general property of quadric surfaces that the line λ meets all the generators l', of which x and t are two particular ones, in related ranges. The line λ, or PS, may evidently be regarded as the intersection of corresponding planes of two related axial pencils of planes, one having x, the other having t, for axis. The point where the line λ meets ϖ, which is the point of contact with one surface of the confocal system of a tangent plane drawn thereto from the line l, may, therefore, be regarded as the intersection of corresponding planes from three related axial pencils of planes, having, respectively, the lines x, t and l for axes. The line of intersection of corresponding planes of two of these axial pencils, as follows from what was said in Chapter I, describes a quadric surface of which both the axes are generators. We thus have, here, three quadric surfaces, say (x, t), (x, l), (t, l), of which any two have a generator in common. It follows that the point where the plane ϖ, through l, meets its corresponding axis, λ, describes a curve which is that common to two quadric surfaces having a common generator. Such a curve evidently meets an arbitrary plane in three points; it is in fact identical with such a cubic curve as that considered above. The theory shews that the curve has two points on each of the lines l, x, t; by similar reasoning it has two points on every polar line, l', of the line l, in regard to the confocal surfaces. And, by definition, it has a simple point on every axis, λ, of a plane ϖ passing through l.

Ex. 1. Putting $b-c=\alpha$, $c-a=\beta$, $a-b=\gamma$, prove that the polar lines of the line whose coordinates are $(l, m, n, \alpha u, \beta v, \gamma w)$, in regard to surfaces confocal with $a^{-1}x^2 + \text{etc.} = t^2$, lie on the quadric surface whose equation is

$$uvw\,(u^{-1}x+v^{-1}y+w^{-1}z)\,(lx+my+nz)-plxt-qmyt-rnzt+lmnt^2=0,$$

where $$p=mv+nw, \quad q=nw+lu, \quad r=lu+mv.$$

Ex. 2. The cubic curve which is the locus of points of contact of tangent planes from this line to the confocals has, for the coordinates of any point,

$$L\,(M^2\gamma-N^2\beta+1), \qquad M\,(N^2\alpha-L^2\gamma+1),$$
$$N(L^2\beta-M^2\alpha+1), \qquad L^2+M^2+N^2,$$

where L, M, N are linear functions of a variable parameter, θ, of which L is given by

$$L=\alpha u\theta+(\gamma mw-\beta nv)\,(\alpha^2u^2+\beta^2v^2+\gamma^2w^2)^{-1},$$

and M, N have similar forms.

When the line l is one of the axes of the confocal system the preceding argument fails. We have already (above, Ex. 2, p. 100) considered the case when the line is an axis in general position. The polars of the line then lie in a plane, and the points of contact of tangent planes to the confocals, drawn through the line, lie on a plane curve having the property of meeting any line of its plane in three points.

When the line l lies in one of the principal planes, its polar lines also lie in a plane, through the corresponding principal point; and the points of contact of the tangent planes, through the line l, lie on a conic in this plane, which meets the principal plane considered in two points conjugate to one another in regard to the focal conic of this plane, the tangents of the locus at these two points meeting in the corresponding principal point. In this case the polar lines of l, in regard to the confocals, all pass through a principal point, that corresponding to the plane in which l lies; by an argument already given, they all meet the axis, λ, in regard to the confocals, of any particular plane, ϖ, drawn through l. Thus these polar lines lie in a plane, σ. The line in which this plane σ meets the principal plane considered is, in fact, the locus of the poles of l in regard to the sections of the confocals by this principal plane. If the plane σ meet l in Q, and O be the pole of l in regard to the focal conic of the principal plane, and P the point of contact of a tangent plane from l to one of the confocals, it is at once seen that the pencils $Q(P)$ and $O(P)$ are related. This proves the statement made.

Lastly, when the line l passes through one of the principal points, the polar lines of l, in regard to the confocals, lie in the corresponding principal plane, as do the axes of the planes drawn through l, all these lines touching a conic, lying in this plane. The circumstances are then sufficiently dealt with by the remarks made above for the case when l is any axis of the confocals.

Ex. When the line l is the line given by $t=0$, $l'x+m'y+n'z=0$, the polar lines of this line in regard to the confocals lie in the plane $\alpha x/l'+\beta y/m'+\gamma z/n'=0$, where $\alpha=b-c$, etc. The points of contact of tangent planes to the confocals drawn from this line lie on the conic expressed by $x=l'(\theta^2+\gamma m'^2-\beta n'^2)$, etc., $t=\theta(l'^2+m'^2+n'^2)$, where θ is a variable parameter.

The enveloping cones drawn from a point to the surfaces of the confocal system, and the tangent planes drawn from a line. We have seen that any plane which touches two surfaces of the confocal system touches every one of these surfaces. Thus, if the two enveloping cones be drawn from an arbitrary point O to two of the surfaces, the four common tangent planes of these two cones, which touch both these surfaces, are equally tangent planes of the cone drawn from O enveloping any other of the surfaces. The enveloping cones from O are thus a system with four common tangent planes. They will meet any one of the principal planes in a system of conics having four common tangent lines, which, in particular, will be tangents of the focal conic of that plane. We may denote the point-pairs in this plane through which these four tangents pass respectively by S_1, H_1; S_2, H_2; S_3, H_3; the three lines S_1H_1, S_2H_2, S_3H_3 will then be a self-conjugate triad in regard to every one of the conics, in particular in regard to the focal conic of the plane. There are thus, through O, three planes forming a self-polar triad for every one of the enveloping cones; and, in regard to any one of the focal conics taken as an absolute conic, these planes are mutually at right angles. The line of intersection of any two of these planes thus contains the pole of the third plane in regard to every one of the focal conics, and, therefore, contains the pole of this plane in regard to all the surfaces of the confocal system. From this it is clear that the three planes are the tangent planes at O of the three confocals of the system which pass through O. Further, a point-pair such as S_1, H_1 is a degenerate form of a conic of the system touching the four lines in the principal plane considered, the tangents of this conic being all lines passing through S_1 or H_1. A degenerate enveloping cone from O consists therefore of all planes passing through one of the lines OS_1, OH_1. Thus these lines are the generating lines at O of one of the confocals which pass through O. The generating lines of the other confocals are, respectively, OS_2, OH_2 and OS_3, OH_3.

The three planes are common *principal* planes of the cones; the six lines OS_1, OH_1, etc., are common *focal* lines of the cones, with reference to the focal conic, taken as absolute conic, in the principal plane considered. As is easy to see, any one of these lines has the property that a plane drawn at right angles to it, in regard to the absolute conic, meets the cone in a conic of which one focus is the intersection of the line with the plane.

Consider now the pairs of tangent planes which can be drawn to the surfaces of the confocal system from an arbitrary line. Taking any point, O, of this line, the pairs of tangent planes from the line to the surfaces are the pairs of tangent planes, from the line, drawn to the enveloping cones of the surfaces which have O for vertex. These cones, however, we have seen, have four common tangent planes; thus, as in the theory of conics in one plane which have four common tangent lines, the pairs of tangent planes to the confocals (two to each surface) are an axial pencil of pairs of planes in involution. There are thus two planes through the line, the double elements of the involution, each of which consists of two coincident tangent planes drawn from the line to one of the surfaces; in other words there are two surfaces of the confocal system which touch the line. And the tangent planes, through the line, of these two confocals, are harmonic conjugates in regard to the pair of tangent planes drawn from the line to any other one of the confocal surfaces. In particular, considering one of the focal conics, and taking it as absolute conic, the tangent planes through the line, of the two confocal surfaces which touch the line, are at right angles to one another in regard to this focal conic. When the line is a generator of one of the confocals, either plane through it touching another confocal, touches all the confocals, and the proof, and theorem, fail.

Ex. 1. Let the equations of the three surfaces of the confocal system, which pass through any point $(x_0, y_0, z_0, 1)$, be

$$(a+\lambda_r)^{-1}x^2+(b+\lambda_r)^{-1}y^2+(c+\lambda_r)^{-1}z^2=t^2,$$

for $r=1, 2, 3$. Also, denote $a+\lambda_r$, $b+\lambda_r$, $c+\lambda_r$, respectively, by a_r, b_r, c_r, and put $p_1{}^2=a_1b_1c_1(a_1-a_2)^{-1}(a_1-a_3)^{-1}$, etc. It may easily be verified that $x_0{}^2(a-b)(a-c)=a_1a_2a_3$, $y_0{}^2(b-c)(b-a)=b_1b_2b_3$, $z_0{}^2(c-a)(c-b)=c_1c_2c_3$, and that $p_1{}^{-2}=a_1{}^{-2}x_0{}^2+b_1{}^{-2}y_0{}^2+c_1{}^{-2}z_0{}^2$, etc. Take for points of reference for coordinates the point $(x_0, y_0, z_0, 1)$, and the points on the plane $t=0$ which lie on the normals, at $(x_0, y_0, z_0, 1)$, of the three confocals through this point; let such coordinates be denoted by $(\xi, \eta, \zeta, 1)$. Prove, then, that the enveloping cone drawn from $(x_0, y_0, z_0, 1)$ to the confocal surface $(a+\lambda)^{-1}x^2+\text{etc.}=t^2$ has for equation

$$(\lambda_1-\lambda)^{-1}\xi^2+(\lambda_2-\lambda)^{-1}\eta^2+(\lambda_3-\lambda)^{-1}\zeta^2=0.$$

Also, that the generators, at $(x_0, y_0, z_0, 1)$, of the three confocals which pass through this point, lie in threes upon four planes, touching all the confocal surfaces, whose equations are

$$(\lambda_2-\lambda_3)^{\frac{1}{2}}\xi\pm(\lambda_3-\lambda_1)^{\frac{1}{2}}\eta\pm(\lambda_1-\lambda_2)^{\frac{1}{2}}\zeta=0.$$

Also, that the polar plane of $(x_0, y_0, z_0, 1)$ in regard to the surface $(a+\lambda)^{-1}x^2+\text{etc.}=t^2$ has for equation

$$(\lambda_1-\lambda)^{-1}p_1\xi+(\lambda_2-\lambda)^{-1}p_2\eta+(\lambda_3-\lambda)^{-1}p_3\zeta-1=0.$$

Denoting this by $U=0$, prove that the equation of the surface $(a+\lambda)^{-1}x^2+\text{etc.}=t^2$ is

$$(\lambda_1-\lambda)^{-1}\xi^2+(\lambda_2-\lambda)^{-1}\eta^2+(\lambda_3-\lambda)^{-1}\zeta^2-MU^2=0,$$

where $M=(\lambda_1-\lambda)(\lambda_2-\lambda)(\lambda_3-\lambda)/(a+\lambda)(b+\lambda)(c+\lambda)$. This fails when λ is one of $\lambda_1, \lambda_2, \lambda_3$; prove that the equation of the surface $(a+\lambda_1)^{-1}x^2+\text{etc.}=t^2$ is capable of either of the two forms

$$(\lambda_1-\lambda_2)^{-1}(p_1\eta-p_2\xi)^2+(\lambda_1-\lambda_3)^{-1}(p_1\zeta-p_3\xi)^2+\xi^2-2p_1\xi=0,$$

$$p_1[N_1\xi^2+(\lambda_1-\lambda_2)^{-1}\eta^2+(\lambda_1-\lambda_3)^{-1}\zeta^2]$$
$$-2\xi[(\lambda_1-\lambda_2)^{-1}\eta p_2+(\lambda_1-\lambda_3)^{-1}\zeta p_3+1]=0,$$

where $N_1=a_1{}^{-1}+b_1{}^{-1}+c_1{}^{-1}-(a_1-a_2)^{-1}-(a_1-a_3)^{-1}, =p_1{}^2\Sigma a_1{}^{-3}x_0{}^2$.

Ex. 2. If we take for points of reference for coordinates the common centre of the confocals and the three points where the normals of the confocals through $(x_0, y_0, z_0, 1)$ meet the plane $t=0$, denoting the coordinates by $(X, Y, Z, 1)$, prove that the sections of the three confocals through $(x_0, y_0, z_0, 1)$ by the planes $X=0$, $Y=0$, $Z=0$ (the planes through the centre parallel, in regard to $t=0$, to the tangent planes at $(x_0, y_0, z_0, 1)$), have the respective equations

$$X=0, \quad (a_1-a_2)^{-1}Y^2+(a_1-a_3)^{-1}Z^2=1;$$
$$Y=0, \quad (a_2-a_1)^{-1}X^2+(a_2-a_3)^{-1}Z^2=1, \text{ etc.}$$

These three conics are therefore such that any plane touching two of them touches the third, and touches the focal conic of the confocals which lies in $t=0$.

Shew that three confocal surfaces can be constructed to have their common centre at $(x_0, y_0, z_0, 1)$, their axes being the normals at this point of the original confocals, to pass through the common centre of the original confocals, and have for normals at this point the axes of the original confocals (cf. Chasles, *Aperçu historique*, 1837, Note xxv, p. 359). Prove, also, that if $(x_0, y_0, z_0, 1)$ lie on a plane drawn through the original centre touching all the original confocals, then this plane equally touches all the new confocals.

Ex. 3. Considering four points P, Q, P', Q' of a line, of which the symbols of P', Q' are expressible by those of P, Q by $P'=P+mQ$, $Q'=P+nQ$, we have previously considered the symbol denoted by $(P', Q'; P, Q)$, which is equal to $m^{-1}n$ (Vol. II, p. 166). We may call this the cross-ratio of P', Q' in regard to P, Q, or of P, Q in regard to P', Q'. Now consider the involution of pairs of points, on the line, of which P, Q are one pair and P', Q' are another pair. Any pair of this involution is given by an equation $\phi+\lambda\psi=0$, where $\phi=0, \psi=0$ are quadratic equations giving two arbitrarily chosen pairs of the involution (Vol. II, p. 111). Any pair of points of the involution is thus given by one parameter, λ. Let λ, λ' be the respective values of this parameter for the two pairs P, Q and P', Q'. Also let w, w' be the respective values of this parameter for the two double pairs of the involution. We may consider the quantity E defined by $E=(\lambda'-w)(\lambda'-w')^{-1}/(\lambda-w)(\lambda-w')^{-1}$, which we may here speak of as the cross-ratio of the parameters λ, λ' in regard to the parameters w, w'. There is a definite relation connecting the cross-ratio $(P', Q'; P, Q)$ with the cross-ratio E. This may be expressed, putting the former equal to $e^{2i\alpha}$, and the latter equal to $e^{2i\beta}$, by $\sin\alpha \sin\beta = \pm 1$. This result can be immediately proved by interpreting the involution on a conic (cf. Vol. II, p. 210). Similar remarks hold of course for involutions of pairs of lines of a plane pencil of lines, or for involutions of pairs of planes of an axial pencil.

We apply this now to the involution of pairs of tangent planes which can be drawn from a line to the surfaces of a confocal system, remarking that the parameter which determines one of these pairs of the involution may be taken identical with that which determines the particular surface to which the planes are tangent. For the pair of tangent planes, drawn from the line of coordinates (l, m, n, l', m', n') to the surface $(a+\lambda)^{-1}x^2 + \text{etc.} = t^2$, is given by $\phi + \lambda\psi = 0$, where $\phi = a\xi^2 + b\eta^2 + c\zeta^2 - \tau^2$, $\psi = \xi^2 + \eta^2 + \zeta^2$, in which $\xi = l't + mz - ny$, etc., $\tau = -(l'x + m'y + n'z)$.

We thus have the result that if, from a line which touches the two confocals of parameters respectively $\lambda = p$, $\lambda = q$, there be drawn the pair of tangent planes to the surface of parameter μ, and also the pair of tangent planes to the surface of parameter ν, then the cross-ratio of the former pair of tangent planes in regard to the latter pair, say $e^{2i\alpha}$, depends upon the cross-ratio of the parameters μ, ν in regard to the parameters p, q, say $e^{2i\beta}$, these being connected by a relation $\sin\alpha \sin\beta = \pm 1$. The former cross-ratio is thus the same as for any other line also touching the two surfaces p, q, proper regard being paid to the order in which the pair of tangent planes to a confocal are taken.

Through any point of the surface of parameter p there can be drawn two lines, lying in the tangent plane of this surface at this point, which also touch the surface of parameter q, these being generators of the enveloping cone from this point to the surface (q). From what has been shewn above in regard to the enveloping cones drawn to the confocals from any point, it is clear that these two lines are a pair of an involution of lines through the point, in the tangent plane, obtained by varying q, of which the double rays are the normals of this point of the other two confocals which pass through the point. It can be shewn that there exist curves on the surface (p) with the property that, the tangent line of the curve, at any point, is a tangent line also of the surface (q); all these curves, which are called *geodesics*, touch the curve in which the surface (p) is met by the surface (q), any tangent line of this latter curve being a common tangent line of the surfaces (p) and (q).

Ex. 4. The two geodesics which can be drawn, as in Ex. 3, from any point of the confocal (p), to touch the curve of intersection, say (p, q), of the surfaces (p) and (q), may suggest the two tangent lines which can be drawn in a plane, to a conic, from an arbitrary point. It can in fact be shewn that, if the various curves, (p, q), of the surface (p), obtained by its intersection with various confocal surfaces (q), be projected from one of the principal points, say $(0, 0, 1, 0)$, each giving rise thereby to a quadric cone, then these cones have four common tangent planes. If we consider on the principal plane $t = 0$, the focal conic on the plane, and the section of the surface (p), there will be two common chords of these conics which intersect in the centre of projection $(0, 0, 1, 0)$; each of these two chords is the intersection of two of the four common tangent planes spoken of. Taking then the focal conic of $t = 0$ as absolute conic, and the two intersections of any plane with this conic as the absolute points of this plane, we reach the conclusion that the curves (p, q'), for various confocals (q'), project from $(0, 0, 1, 0)$, upon any plane of circular section of the surface (p) which does not contain this point, into a system of confocal conics, $[p, q']$. For definiteness consider one of the planes of circular section which contains the common centre of all the confocals, and also the point $(0, 1, 0, 0)$. Two foci of the confocal conics are then projections of umbilici in the plane $y = 0$.

Now consider a particular point of the surface (p), say P, and the two tangent lines through P which also touch the surface (q). These are the intersections with the tangent plane at P of the two tangent planes drawn to the surface (q) from the normal, at P, of the surface (p). The cross-ratio of these planes in regard to the two tangent planes drawn from this normal to the focal conic of the plane $t = 0$ (which touch this conic at points lying on the tangent plane at P) depends, we have seen, upon the cross-ratio of the two parameters, q and ∞, giving these confocals, in regard to the two parameters, q_1 and q_2, of the two confocals, other than (p), which pass through P, namely depends on $(q - q_1)/(q - q_2)$. Again, the point $(0, 0, 1, 0)$ being C, as the pairs of tangent planes from CP, to the cones projecting from C the curves (p, q'), are in involution, the cross-ratio of the two tangent lines from $[P]$, the projection of P, to the conic $[p, q]$, in regard to the two lines joining $[P]$ to the absolute points of the plane of projection, depends in the same way upon the same quantity $(q - q_1)/(q - q_2)$. Thus, defining the measure of angle as in Vol. II, pp. 167, 192, we may say that the two tangents from P to the surface (q) intersect at the same angle as the two tangents from $[P]$ to the conic $[p, q]$. The two former lines however do not project into the two latter.

Ex. 5. Consider the curve of intersection of two confocals $(a + \mu)^{-1}x^2 + \text{etc.} = t^2$, $(a + \nu)^{-1}x^2 + \text{etc.} = t^2$, say the surfaces (μ), (ν). Through any point, P, of this curve there pass two generators of the third confocal, say (λ), which can be drawn through P. Consider one of these two generators, say p. Through any other point P', of this generator, there pass, beside the surface (λ), two other confocals, (μ') and (ν'), and the curve of intersection of these, say the curve (μ', ν'). Consider all these curves (μ', ν'), for varying positions of P' on this chosen generator p. They are all intersected by one of the two generators of the third confocal which can be drawn through any other point, Q, of the first curve (μ, ν), say the generator q. The generators, q, so obtained meet any one of the principal planes in the points of a conic. Also, if the generator q meet the curve (μ', ν') in the point Q', the range of points Q' on q, as the curve (μ', ν') varies, but so as always to meet the generator p, in P', is related to the range of points P' on the generator p. A particular case is that the two points, say Q_1 and Q_2, in which any particular confocal (M), other than (λ), is met by the generator q, each describes, on this confocal (M), as Q varies on the curve (μ, ν), a curve, (M, N), which is the intersection of the surface (M) with another confocal (N). Thus, if the surface (M) be met by p in P_1 and P_2, and Q', Q'' be two positions of Q' corresponding to the positions P', P'', of P', the cross-ratio $(Q', Q''; Q_1, Q_2)$ is equal to $(P', P''; P_1, P_2)$. As in the case of a plane, where we may measure the interval between two points with reference to an absolute conic, we may thence speak of the interval (Q', Q''), with reference to the surface (M), as being equal to the interval (P', P''). But, also as in the case of a plane, we may replace (M) by one of the focal conics, and so obtain a Euclidean measure of distance (cf. Vol. II, p. 184). In this sense we may speak of the two generators p, q as being divided into segments of equal length by the curves (μ', ν'). The measurement of distance, and the theory of movements, will be considered in a later volume.

The analytical proof of what has been stated is very easy. It is easy to verify that one of the eight points of intersection of the surfaces (λ), (μ), (ν), say, the point P, has coordinates

$$\left[\frac{(a+\lambda)(a+\mu)(a+\nu)}{(a-b)(a-c)}\right]^{\frac{1}{2}}, \quad \left[\frac{(b+\lambda)(b+\mu)(b+\nu)}{(b-c)(b-a)}\right]^{\frac{1}{2}},$$

$$\left[\frac{(c+\lambda)(c+\mu)(c+\nu)}{(c-a)(c-b)}\right]^{\frac{1}{2}}, \qquad 1,$$

or say $[(a+\lambda)^{\frac{1}{2}} U,\ (b+\lambda)^{\frac{1}{2}} V,\ (c+\lambda)^{\frac{1}{2}} W,\ 1]$. Consider now the point in the plane $t=0$ whose coordinates are

$$[(a+\lambda)^{\frac{1}{2}} u,\quad (b+\lambda)^{\frac{1}{2}} v,\quad (c+\lambda)^{\frac{1}{2}} w,\quad 0],$$

where

$$u=[(a+\lambda')/(a-b)(a-c)]^{\frac{1}{2}},\quad v=[(b+\lambda')/(b-c)(b-a)]^{\frac{1}{2}},\ \text{etc.},$$

say the point T. Here λ' is to be so taken that

$$(b-c)[(a+\mu)(a+\nu)(a+\lambda')]^{\frac{1}{2}}+(c-a)[(b+\mu)(b+\nu)(b+\lambda')]^{\frac{1}{2}} \\ +(a-b)[(c+\mu)(c+\nu)(c+\lambda')]^{\frac{1}{2}}=0,$$

a condition which, for given values of μ and ν, is satisfied by two values of λ'. Thence, in addition to the identity, $u^2+v^2+w^2=0$, we have the relation $Uu+Vv+Ww=0$. We also have the identity $U^2+V^2+W^2=1$. Now consider, for any value of h, the point of coordinates

$$(a+\lambda)^{\frac{1}{2}}[U+hu],\quad (b+\lambda)^{\frac{1}{2}}[V+hv],\quad (c+\lambda)^{\frac{1}{2}}[W+hw],\quad 1;$$

for all values of h this point lies on the surface $(a+\lambda)^{-1}x^2+\text{etc.}=t^2$; for fixed values of λ, μ, ν, and one of the two corresponding fixed values of λ', this point describes a line as h varies. The point is, therefore, the general point of one of the two generators above explained, at P, of the surface (λ). It meets the plane $t=0$ in the point T, which, as λ varies, μ, ν and therefore λ' remaining fixed, describes the conic $(a+\lambda')^{-1}x^2+\text{etc.}=0$, $t=0$. The line similarly meets the plane $x=0$ in the point of coordinates $(0, -w, v, iu)$, for we easily see that $vW-wV=iu$, $wU-uW=iv$, etc. As λ varies this point describes the conic $(b+\sigma)^{-1}y^2+(c+\sigma)^{-1}z^2=t^2$, $x=0$, where $\sigma=(bv^2+cw^2)u^{-2}$.

To find the curve (μ', ν') which passes through any point of the generator p, we have to solve for μ', ν' the three equations such as

$$(a+\lambda)^{\frac{1}{2}}(U+hu)=(a+\lambda)^{\frac{1}{2}}U',$$

wherein U' denotes $[(a+\mu')(a+\nu')/(a-b)(a-c)]^{\frac{1}{2}}$. Herein λ, μ, ν, h are given, and λ' is that, of the two possible values, which belongs to the generator p, through P, the other value of λ' belonging to the other generator of the surface (λ) through this point. The three equations are equivalent to two, as we see from $U'^2+V'^2+W'^2=1$, and lead to definite values of μ', ν'. These values are however independent of λ; thus as λ, and the surface (λ), vary, the curves (μ', ν') are all met by generators, q, meeting the curve (μ, ν), each lying on one of the surfaces (λ). That the ranges on the generators q, by the various curves (μ', ν'), are related, is clear from the fact that the value of h which belongs to a point P' on p is the same as that belonging to the point Q' of q which lies on the same curve (μ', ν'). Incidentally this argument shews that the lines q meet any one of the confocals, say (M), upon two curves (μ', ν'), say (M, N_1) and (M, N_2). In particular, to find the locus of the other intersection, beside the curve (μ, ν), of the lines q with the surface (μ), we shall be required to solve, for ν_1, the equations

$$(a+\lambda)^{\frac{1}{2}}(U+hu)=(a+\lambda)^{\frac{1}{2}}U_1,$$

where

$$U_1=[(a+\mu)(a+\nu_1)/(a-b)(a-c)]^{\frac{1}{2}}, \text{ etc.}$$

Beside $\Sigma U_1^2=1$, these equations require $\Sigma(a+\mu)^{-1}U_1^2=0$, which determines h; then ν_1 is to be found from

$$\Sigma U_1 u=0, \quad \text{or} \quad \Sigma(b-c)[(a+\mu)(a+\lambda')(a+\nu_1)]^{\frac{1}{2}}=0;$$

of the two roots of this equation for ν_1, one is ν itself. The remainder of the statement made is now clear.

The equation above, by which λ' is determined from μ and ν, is capable of interpretation by elliptic functions, and will be considered in a later section. Geometrically, it expresses that one of the intersections of the three confocals (μ), (ν), (λ') lies on one of the four planes drawn from the common centre of the confocals to touch all these surfaces, the equation of such a plane being $\Sigma x(b-c)^{\frac{1}{2}}=0$.

8—2

Thus, if λ_1', λ_2' denote the two values of λ' when μ and ν are given, we infer that each of these four planes contains two of the intersections of (μ), (ν), (λ_1'); and two of the intersections of (μ), (ν), (λ_2').

If the equation connecting μ, ν and λ' be rationalised it will be found to be symmetrical in regard to the two sets μ, ν, λ' and a, b, c. Another irrational form of the equation is therefore

$$(\mu - \nu) L' + (\nu - \lambda') M + (\lambda' - \mu) N = 0,$$

where

$$L' = (a + \lambda')^{\frac{1}{2}} (b + \lambda')^{\frac{1}{2}} (c + \lambda')^{\frac{1}{2}}, \text{ etc.}$$

Further, denoting the equation by $(\mu, \nu, \lambda') = 0$, it can be shewn that λ, μ, ν can be eliminated from the three equations $(\mu, \nu, \lambda') = 0$, $(\nu, \lambda, \mu') = 0$, $(\lambda, \mu, \nu') = 0$, the result being $(\lambda', \mu', \nu') = 0$.

Ex. 6. From the preceding results we have an immediate proof of the porism of Poncelet[1] in regard to conics (Vol. II, p. 55), so far as it relates to triangles. Here it may be stated in the form: If the join of a point, M, lying on the conic $(a + \mu)^{-1} x^2 + \text{etc.} = 0$, to a point, N, lying on the conic $(a + \nu)^{-1} x^2 + \text{etc.} = 0$, be a tangent of the conic $x^2 + y^2 + z^2 = 0$, the locus of the point of intersection, T, of the other tangents drawn to this latter conic from the points M, N is a third conic, $(a + \lambda')^{-1} x^2 + \text{etc.} = 0$. The four conics involved all have four common tangents. To prove this, consider, in the first instance, any three confocals of the system of confocal quadrics, (λ), (μ), (ν), intersecting in a point O. Let the generators at O, of the surface (λ), meet the plane $t = 0$ in the points T, T_1, lying on the conic $(a + \lambda)^{-1} x^2 + \text{etc.} = 0$, $t = 0$; the generators at O, of the surface (μ), meet the plane $t = 0$ in the points M, M_1, lying on the conic $(a + \mu)^{-1} x^2 + \text{etc.} = 0$; and the generators at O, of the surface (ν), meet the plane $t = 0$ in the points N, N_1, lying on the conic $(a + \nu)^{-1} x^2 + \text{etc.} = 0$. By what we have seen (p. 110), the six points T, T_1, M, M_1, N, N_1 lie in threes on four lines, in which the plane $t = 0$ is met by the four planes drawn from O to touch all the confocals; we suppose the notation so chosen that these lines are MNT_1, TMN_1, TNM_1, $T_1M_1N_1$; these lines all touch the conic $x^2 + y^2 + z^2 = 0$, $t = 0$. If the line joining O to the common centre, D, of the confocals, meet the plane $t = 0$ in H, the polar line of DH, in regard to any of the surfaces (λ), (μ), (ν), lies in the plane $t = 0$, and, therefore, contains the points of contact of the tangents drawn from H to the section of that confocal by $t = 0$; thus these points of contact lie on the tangent plane at O of the corresponding confocal. In other words, the six tangents of the conics at T, T_1, M, M_1, N, N_1 meet in H. Now, retaining the surfaces (μ), (ν) the same, allow the surface (λ) to vary, the point O varying on the curve (μ, ν). Then

[1] See frontispiece of this Volume.

the points M, N will vary on the conics (μ), (ν), the line MN continuing to touch the focal conic, and, by what is proved above, the point T, the intersection with $t=0$ of one of the generators at O of the surface (λ), will describe a conic (λ'), where λ' is connected with μ, ν by the relation $(\mu, \nu, \lambda')=0$. The other value of λ' which satisfies this equation gives the conic (λ_1'), described by the point T_1, where the other generator at O of the surface (λ) meets $t=0$.

Conversely, if, as above, M, N be two points of the plane $t=0$, lying respectively on the conics (μ), (ν), the line MN touching the focal conic, and the other tangents to this conic from M and N intersecting in T, we can construct the figure above used, and the locus of T, as follows: By hypothesis the conics (μ), (ν) on which M and N lie, and the focal conic touched by MN, have four common tangents. We can therefore refer them to a common self-polar triad, and suppose their equations to be as we have taken them above, and then, further, taking an arbitrary point D, not in the plane $t=0$, suppose these conics to arise from a confocal system of quadrics, as above, with D as centre. Then, through each of the lines TM, MN, NT there pass two planes which touch all the surfaces of this confocal system, a pair of such planes having equations of the form

$$lx+my+nz\pm pt=0, \quad \text{where} \quad p^2=al^2+bm^2+cn^2.$$

The three pairs of planes give rise to eight points through each of which passes a plane of each pair, these eight points lying in couples on four lines through D. Taking one of these eight points for the point O, such a figure as that above described can be constructed; but, the conics (μ), (ν) being given, and, therefore, the point, H, of intersection of the tangents of these conics at M and N, it is necessary to take for O one of the two points such that DO passes through H. These two points O give rise to the same surface (λ), and the same conic (λ'), the point T being the intersection of two generators of the surface (λ).

When M, N vary, respectively on the conics (μ), (ν), the conic (λ) varies, and T remains on (λ'), while H describes a conic, the projection from D of the curve (μ, ν). The conic (λ'), and the conic (λ), have, at T, tangents which are conjugate in regard to the focal conic of the plane $t=0$. We may similarly consider the conics, (μ'), through M, and (ν'), through N, whose tangents at these points are conjugate, respectively, to those of (μ) and (ν). Then, when M and N vary on (μ'), (ν'), respectively, T also varies on (λ'). Calling for brevity the line conjugate, in regard to the focal conic, to a tangent of one of the conics, the *normal*, we may thus say that when a triad of points, one on each of the conics, is such that their joins touch the focal conic, not only the tangents of two of these conics and the normal of the other, meet in a point, but also the normals of these conics meet in a point, and these points of concurrence describe conics.

Ex. 7. If the line joining two points P_0, P, of respective coordinates $(x_0, y_0, z_0, 1)$, $(x, y, z, 1)$, meet the confocal surface $(a+\theta)^{-1}x^2 + \text{etc.} = t^2$ in the points M, N, and the cross-ratio $(P_0, P; M, N)$ be written e^{2iw}, prove that

$$\cosh w = \frac{(\theta+a)(\theta+b)(\theta+c) - \Sigma x x_0 (\theta+b)(\theta+c)}{[\Pi(\theta-\lambda)(\theta-\lambda_0)]^{\frac{1}{2}}},$$

where λ_0, μ_0, ν_0 are the parameters of the confocals through P_0, and λ, μ, ν the parameters of the confocals through P, so that we may replace x_0 by $[(a+\lambda_0)(a+\mu_0)(a+\nu_0)/(a-b)(a-c)]^{\frac{1}{2}}$, and x by the same function of λ, μ, ν. This expression is unaffected by the interchange of λ and λ_0, whereby P_0 is replaced by a point P_0' on the same curve (μ_0, ν_0), and P is replaced by a point P' on the same curve (μ, ν), provided the signs of xx_0, yy_0, zz_0 are unaltered.

The value of $\cosh^2 w$ is unity, not only for $\theta = -a, -b, -c, \infty$, but also for two other values of θ; these give the two confocals touched by the line P_0P. Thus the line $P_0'P'$ also touches these two confocals. And the cross-ratio $(P_0, P; M, N)$ is equal to that obtained, for the same surface θ, for the line $P_0'P'$.

By supposing θ to become infinitely great, and regarding the focal conic of the plane $t=0$ as the absolute conic, it can hence be inferred, as will be seen later, that the *distances* P_0P and $P_0'P'$, which arise as the limits of $\theta^{\frac{1}{2}}w$, are equal. This result is known as Ivory's theorem, and is of importance in the theory of the Newtonian attraction of a material ellipsoid. The geometrical theory is capable of great development. See a paper by A. L. Dixon, *Messenger of Mathematics*, XXXII, 1903, p. 177, and a note by the author, *Proc. Camb. Phil. Soc.* XX, 1920, p. 129.

Ex. 8. The planes touching all the confocals of the system are those represented by the equation $Px + Qy + Rz = t$, where

$$P = (a+\lambda)^{\frac{1}{2}}(a-b)^{-\frac{1}{2}}(a-c)^{-\frac{1}{2}}, \text{ etc.}$$

These planes, depending on a single parameter, determine a developable. They meet an arbitrary plane in the tangent lines of a curve. Upon each plane, of the developable, there is a line, of which the points are given by $x = t(a+\phi)P$, etc., for varying values of ϕ; the intersections of these lines with the arbitrary plane are the points of the curve spoken of. The planes of the developable meet, each in three coincident points, a curve of which the general point is given by $x = t(a+\lambda)^{\frac{3}{2}}(a-b)^{-\frac{1}{2}}(a-c)^{-\frac{1}{2}}$, etc. The lines in question are called the generating lines of the developable; each of them touches all the confocals, other than one of these, of which it is a generator; through this line there is one plane of the developable touching all the confocals at points of this line. More generally, the line of intersection of any two planes of the developable is a generator of one of the confocal surfaces; for the surface drawn to touch an arbitrary plane through this line of intersection will be touched by three planes through the line; it will thus contain the line entirely. If we call the line of intersection of two planes of the developable an *axis* of the developable, (it corresponds dually to a chord of a curve), we thus see that the aggregate of all the generators of the confocals is the same as the aggregate of the axes and generating lines of the developable.

The line given by $x = t(a + \phi) P$, etc., is a generator of the surface $(a + \lambda)^{-1}x^2 + \text{etc.} = t^2$, touching the surface $(a + \theta)^{-1}x^2 + \text{etc.} = t^2$ at the point given by $\phi = \theta$; this point is one of the eight intersections of the surfaces (λ), (θ) with the quadric which is the polar reciprocal of the surface (λ), on which the line is a generator, with respect to the surface (θ). (Cf. pp. 80, 92.)

Dual method of developing the theory of confocal surfaces. If any two quadric surfaces be given, the curve in which they intersect is such as to have four points upon an arbitrary plane, these being the intersections of the conics in which the plane severally meets the quadrics. Through this curve, and an arbitrary point, can be drawn another quadric; for if $S = 0$, $S' = 0$ be the equations of the given quadrics, the equation $S + \lambda S' = 0$ represents a quadric passing through the common points of the two given quadrics; and the value of λ can be taken so that this passes through the arbitrary point. No other quadric can be drawn through this curve and this point, since, of such a quadric, five points are determined lying on an arbitrary plane drawn through this point. Thus the equation $S + \lambda S' = 0$ is that of the general quadric through the curve $S = 0$, $S' = 0$. We have seen that the tangential equation of the general quadric of the confocal system is of the form $\Sigma + \lambda\Sigma' = 0$; we may, then, clearly develop, in dual form, the theory of a confocal system, as that of the quadrics $S + \lambda S' = 0$. Without entering into a detailed account, it is interesting to consider what corresponds to some of the leading theorems already obtained for confocals. We suppose that the surfaces $S = 0$, $S' = 0$ have a common self-polar tetrad, so that their equations can be supposed to be, respectively,

$$Ax^2 + By^2 + Cz^2 + Dt^2 = 0, \quad A'x^2 + \text{etc.} = 0;$$

their curve of intersection will then not contain a line, or a conic, as part, as is possible for two quadrics in general. Through the curve of intersection there pass four quadric cones; in fact $S + \lambda S' = 0$ represents a cone for $\lambda = -A/A'$, etc. These cones, each enveloped by an aggregate of planes passing through one of the four principal points, of which two planes can be drawn to contain an arbitrary line through this point, correspond dually to the four focal conics of the confocal system, each the locus of points in a principal plane, of which two points lie on an arbitrary line of the plane. The coordinates can be chosen so that the equation of the system of quadrics now considered is

$$(a+\mu)x^2+(b+\mu)y^2+(c+\mu)z^2=t^2,$$

where $\mu=\infty$ gives rise to one of the cones, $x^2+y^2+z^2=0$. The other three cones then arise for $\mu=-a$, $\mu=-b$, $\mu=-c$. It is at once clear that the polar planes of an arbitrary point, in regard to the system of quadrics now being considered, all pass through a line. These polar planes have an equation $P+\mu P_0=0$, where P, P_0 are the polar planes of the point in regard to the fixed quadrics $ax^2+\text{etc.}=t^2$, and $x^2+y^2+z^2=0$. The line is the intersection of four planes, arising for $\mu=\infty, -a, -b, -c$, each containing one of the principal points, and the axial pencil formed by these planes is related to that similarly arising for any other such line. These are the lines corresponding to what have been called the axes of the confocal system; as in that case they constitute a tetrahedral complex; every line in a principal plane, or through a principal point, belongs to this complex. Considering, next, an arbitrary line, this is intersected, by the quadrics considered, in pairs of points belonging to an involution, the two points of intersection with any quadric being determined by one of them. There are thus two quadrics of the system touching the line. The proof and theorem fail if the line contains two points of the common curve of intersection of the quadrics, in which case the line is a generator of one of the quadrics; when the line is a tangent of this curve, it is a generator of one of the quadrics, and touches all the other quadrics at this point. In the general case, the tangent planes of the quadrics, at the points where the line meets them, form an aggregate of which three pass through an arbitrary point; also the polar lines of the given line in regard to the quadrics are generators of a quadric, which passes through the four principal points. When the line belongs to the tetrahedral complex, the tangent planes of the quadrics, at the point where the line meets them, all pass through a point. In particular, when the line passes through one of the principal points, these tangent planes touch a quadric cone, whose vertex is in the corresponding plane. Lastly, considering an arbitrary plane, there are three quadrics of the system which touch this plane. The quadrics of the system meet this plane in conics having four points in common; the generators, in this plane, of the three quadrics which touch this plane, are the three line-pairs which contain these four points. If we consider the aggregate of planes touching two particular quadrics of the system, and, in any such plane, the generators of the third quadric of the system other than these two, which touches this plane, then these generators are projected from one of the principal points by tangent planes of two particular quadric cones.

Ex. 1. The polar lines of the line (l, m, n, l', m', n'), in regard to the quadrics of equation $(a+\lambda)x^2 + \text{etc.} = t^2$, lie on the quadric expressed by

$$l'\alpha yz + m'\beta zx + n'\gamma xy - lxt - myt - nzt = 0,$$

where $\alpha = b - c$, $\beta = c - a$, $\gamma = a - b$. When the line belongs to the tetrahedral complex expressed by $all' + bmm' + cnn' = 0$, this quadric is a cone of vertex $(l', m', n', -\sigma)$, where σ is $\alpha m'n'/l, = \beta n'l'/m$, etc. When the line passes through one of the principal points the quadric breaks up into two planes; for instance, when $l' = m' = n' = 0$, it is $t(lx + my + nz) = 0$.

Ex. 2. The tangent planes of the quadrics $(a+\mu)x^2 + \text{etc.} = t^2$, at the points where these are met by the line $x/l = y/m = z/n$, touch the cone

$$(xP + yQ + zR)(xp + yq + zr) = t^2,$$

where $(l^2 + m^2 + n^2)P = l(\gamma m^2 - \beta n^2)$, $(l^2 + m^2 + n^2)p = l$, etc., so that $Pp + Qq + Rr = 0$.

CHAPTER III

CUBIC CURVES IN SPACE. THE INTERSECTION OF TWO OR MORE QUADRICS

The curve of intersection of two quadrics. A conic is a curve of which the points are in unique correspondence with the points of a line; it is defined by the intersection of associated rays of two related flat pencils of lines in a plane. In three dimensions there exists a curve, in many ways simpler than a conic, of which the points are also in unique correspondence with the points of a line. Three points of this curve lie on an arbitrary plane, so that the curve is called a cubic. The curve may be defined, analogously to a conic in a plane, by the intersection of associated planes of three related axial pencils of planes. The curve may also be defined as the partial intersection of two quadric surfaces. We have, however, in the theory of confocal quadrics, been led to consider a curve which is the intersection of two quadrics, and is not a cubic. It will therefore add to clearness if, before considering the cubic curve in detail, we enumerate all the possibilities in regard to the intersection of two quadrics; the proof that the list is exhaustive will be given in a later Volume. Since an arbitrary plane meets each of the quadrics in a conic, and two conics in the same plane intersect in four points, the complete intersection of two quadrics must meet an arbitrary plane in four points. This intersection may however consist of several irreducible pieces (curves), of which two may coincide.

(1) The curve of intersection of two quadrics may consist of one piece, that is of one curve, which will then be a quartic, as meeting an arbitrary plane in four points. In the most general case the points of this curve are not in unique correspondence with the points of a line. If we take two points of the curve, and the line, or *chord* of the curve, which joins them, and consider the planes through this chord, (which are in correspondence with the points of a line), each of these planes gives rise to two points of the curve. The coordinates of the points of the curve are in fact rational functions of two parameters, ξ, η, themselves connected by a relation $\eta^2 = f(\xi)$, wherein $f(\xi)$ is a polynomial in ξ of the fourth order; conversely, both ξ and η are rational functions of the ratios of the coordinates of a point of the quartic curve.

There is however a particular case of this in which the curve of intersection still consists of a single quartic curve; namely when the curve crosses itself, or has a double point. Then any plane drawn through this point and another fixed point of the curve has but one further intersection with the curve; the points of the curve are in this case in unique correspondence with the points of a line. The cases in which the curve crosses itself twice occur below; the curve does not then consist of one piece.

(2) The curve may consist of a line and a curve consisting of one piece. This last is then the cubic curve above referred to; it meets the line in two points.

(3) The curve may consist of two conics, which then have two points of intersection. We have met with this possibility in the last chapter, in the case of two spheres, one of the common conics being then the absolute conic. The two conics may touch one another.

(4) In particular one of the conics may consist of two straight lines, each of which meets the other conic. When the point of intersection of these lines lies on the other conic, the plane of the lines contains the tangent line of this conic at this point.

(5) Or both conics may consist of two lines. In this case the two quadrics have in common two generators of each of the two systems of either quadric.

(6) Two of the four lines may coincide, in which case the intersection consists of a line at every point of which the two quadrics have the same tangent plane, together with two lines intersecting this.

(7) More particularly still, the four lines may consist of two, each occurring twice; the lines will intersect, and the tangent planes of the quadrics will be the same at every point of both.

(8) Finally, the intersection may consist of a single conic at every point of which the quadrics have the same tangent plane. This conic thus counts doubly in the intersection.

In the cases (6), (7), (8) we say that the common curve is multiple, or contains a multiple portion.

Ex. 1. The quadrics whose equations are $x^2+y^2-z^2-t^2=0$, $ax^2+by^2+cz^2+dt^2=0$, intersect in a curve of which the points are expressed by $x=\theta+\phi$, $y=1-\theta\phi$, $z=1+\theta\phi$, $t=\theta-\phi$, provided θ, ϕ are connected by the equation

$$A(1+\theta^2\phi^2)+B(\theta^2+\phi^2)+2C\theta\phi=0,$$

where $A=b+c$, $B=a+d$, $C=a-b+c-d$.

The quadrics whose equations are

$$2xy+z^2-t^2=0,\quad 2xy+y^2+(1-m^2)z^2-(1-n^2)t^2=0$$

intersect in a curve expressed by

$$x=PQ,\quad y=2m^2n^2,\quad z=mn^2(\theta+\theta^{-1}),\quad t=m^2n(\theta-\theta^{-1}),$$

where

$$2P = (\theta - \theta^{-1})\, m + (\theta + \theta^{-1})\, n, \quad 2Q = (\theta - \theta^{-1})\, m - (\theta + \theta^{-1})\, n,$$

θ being variable. This curve crosses itself at (0, 1, 0, 0), either of the two planes, $mz + nt - \theta y = 0$, $mz - nt - \theta^{-1} y = 0$, meeting the curve in only one further point.

The quadrics whose equations are

$$2xy + z^2 + t^2 = 0, \quad 2xy + 2yz + z^2 + (1 - n^2)\, t^2 = 0$$

intersect in a curve expressed by

$$x = -(\theta^2 + n^4), \quad y = \tfrac{1}{2}\theta^4, \quad z = n^2\theta^2, \quad t = \theta^3,$$

where θ varies. This curve has a cusp at (0, 1, 0, 0), any plane $2y = \theta t$ meeting the curve in only one further point.

Ex. 2. The quadrics whose equations are

$$xy + zt = 0, \quad 2xy + 2czt + y^2 + t^2 = 0$$

intersect in a curve expressed by

$$x = -(1 + \theta^2), \quad y = 2(1 - c)\,\theta^2, \quad z = \theta(1 + \theta^2), \quad t = 2(1 - c)\,\theta,$$

as well as in the line $y = 0$, $t = 0$. The curve meets an arbitrary plane in three points, and meets the line in (1, 0, 0, 0) and (0, 0, 1, 0).

The quadrics whose equations are

$$xy + zt = 0, \quad 2xy + 2zt + 2yt + z^2 = 0$$

intersect in a curve expressed by $x = 1$, $y = -2\theta^3$, $z = 2\theta^2$, $t = \theta$, as well as in the line $y = 0$, $z = 0$. This line meets the curve in two coincident points, any plane $y + \theta z = 0$ meeting the curve in only one further point.

Ex. 3. The quadrics $x^2 + y^2 + z^2 + t^2 = 0$, $x^2 + y^2 + cz^2 + dt^2 = 0$ meet in two conics having two points in common. The quadrics $2xy + z^2 + t^2 = 0$, $2xy + y^2 + z^2 + dt^2 = 0$ meet in two conics touching one another on the line $y = 0$, $t = 0$.

Ex. 4. The quadrics $2xy + z^2 - t^2 = 0$, $2xy + y^2 + c(z^2 - t^2) = 0$ intersect in two lines $y = 0$, $z = \pm t$, and in a conic, lying in the plane $y + 2x(1 - c) = 0$, meeting both these lines. The quadrics $2xy + z^2 - t^2 = 0$, $2xy + 2yz + z^2 - t^2 = 0$ intersect in two lines $y = 0$, $z = \pm t$, and in a conic, lying in the plane $z = 0$, which contains the intersection of the two lines and is touched, at this point, by the plane of these.

Ex. 5. The quadrics $x^2 + y^2 + z^2 + t^2 = 0$, $x^2 + y^2 + c(z^2 + t^2) = 0$ have four lines in common.

Ex. 6. The quadrics $xy + zt = 0$, $xy + zt + yt = 0$ have in common the lines $x = 0 = t$, $y = 0 = z$, and $y = 0 = t$, of which the last counts doubly, the tangent planes of the quadrics at any point of this being the same.

Ex. 7. The quadrics $xy + zt = 0$, $2xy + 2zt + y^2 = 0$ have in common the lines $y = 0 = t$ and $y = 0 = z$, each doubly, the tangent planes of the quadrics at any point of either being the same.

Ex. 8. The quadrics $x^2 + y^2 + z^2 + t^2 = 0$, $x^2 + y^2 + z^2 + dt^2 = 0$ touch one another at every point of the conic $t = 0$, $x^2 + y^2 + z^2 = 0$, and have no other common point.

Ex. 9. We have seen that the theory dual to that of the theory of confocal surfaces is that of the quadrics passing through the common curve of two quadrics of sufficient generality. In particular, the dual of a focal conic, of which every tangent line lies in (two of the) planes touching all the confocal surfaces, is a quadric cone of which every generating line passes through a point of the common curve of the two quadrics. It is thus of interest to determine, in each of the thirteen cases just enumerated, the quadric cones containing the complete curve of intersection of the two quadrics. Moreover, the tangent planes touching the confocal surfaces are the common tangent planes of any two of the focal conics. It is thus of interest to consider the aggregate of tangent planes common to two conics which are not in so general relation to one another as are two of the focal conics; in particular, of two conics in different planes which both touch the line of intersection of their planes. It is this problem, essentially, which we now proceed to consider, in its dual form.

The cubic curve in three dimensions. Consider first two quadric cones, of vertices O and Q, so related that the line OQ is a generator of both, the tangent planes of the cones along this generator being different. These cones meet an arbitrary plane in two conics, among the four intersections of which will be the point in which the line OQ meets the plane. There are then three points of the plane, not on this line, which belong to both cones. The curve of points common to the two cones, other than their common generator, thus meets an arbitrary plane in three points, and is, for this reason, called a *cubic* curve. Any plane through OQ meets each cone in a further generator; the point of intersection of these two lines is a point of the curve; the other two points of the curve lying in this plane are the points O and Q. If R be any point of the curve, any plane drawn through R meets the curve in two other points; if then we consider the aggregate of lines joining R to all other points of the curve, there are two of these in any plane through R. We infer, therefore, that these lines are the generators of a quadric cone. This may be deduced also by remarking that the axial pencils of planes which join two fixed generators of a quadric cone to a variable generator are related pencils, and conversely. For let P be a variable point of the curve; as OP is a variable generator of the quadric cone of vertex O, of which OQ and OR

are two fixed generators, the axial pencils $OQ(P)$ and $OR(P)$ are related; also, as QP is a variable generator of the quadric cone of vertex Q, of which QO and QR are two fixed generators, the axial pencils $QO(P)$ and $QR(P)$ are related. As then the axial pencils $RO(P)$ and $RQ(P)$ are related, it follows that RP is the variable generator of a quadric cone of vertex R, of which RO and RQ are two fixed generators. The cubic curve may then be defined as the locus of the common points of the two quadric cones with vertices at O and R, other than their common generator OR. It follows too that the axial pencil of planes, joining the chord OQ to the points P of the curve, is related to the axial pencil joining the chord QR to the same points P, and, therefore, to the axial pencil joining any other chord, RS, to the same points P of the curve. In view of what has been said previously in regard to related ranges, this is sufficiently clear also from the fact that an arbitrary plane through OQ, by determining a definite further point, P, of the curve, determines a definite plane RSP, and conversely. The curve is thus like a conic in being the seat of related ranges, of points lying upon it, any four points, P_1, P_2, P_3, P_4, of the curve, being related to four other points, P_1', P_2', P_3', P_4', if the planes joining the first four points, to an arbitrary chord of the curve, be related to the planes joining the second four points to another, also arbitrary, chord. In other words, the curve, like a conic, may be regarded as having its points in (1, 1) correspondence with the points of a line. The curve is, in fact, as appears from what has been said, the locus of the intersection of three corresponding planes; one from each of three related axial pencils of planes, of which the axes are any three chords of the curve.

Consider any two points D, A, of the curve, and the quadric cones which project the curve from these two points, respectively, say (D) and (A). Let the tangent plane of the cone (D) along DA meet the cone (A) also in the line AB, and the tangent plane of the cone (A) along AD meet the cone (D) also in the line DC. Let B be determined as the intersection of the tangent plane of (D), along DC, with AB; and C be determined as the intersection of the tangent plane of (A), along AB, with DC. Then, with A, B, C, D as fundamental points for coordinates, the cones (D), (A) may be supposed to have the respective equations $xz = y^2$ and $yt = z^2$. The former is generated by the intersection of planes of equations $x = y\theta$, $y = z\theta$; the latter by the intersection of planes of equations $y = z\theta$, $z = t\theta$, where θ is variable. The coordinates of any point of the curve are then of the form $(\theta^3, \theta^2, \theta, 1)$. The three points in which any plane, of equation $lx + my + nz + pt = 0$, meets the curve, are then given by the roots of the cubic equation $l\theta^3 + m\theta^2 + n\theta + p = 0$. In particular, the plane $x = 0$, or BCD, meets the curve in three coincident points at the point D, for which $\theta=0$; and the plane $t=0$, or BCA, meets the curve in three coincident points at the point A, for which $\theta^{-1}=0$. The line DC meets the curve in two coincident points at the point D, any plane, $x=y\theta$, through this line, meeting the curve in only one other point; so the line AB meets the curve in two coincident points at the point A, any plane, $z=t\theta$, through this line, meeting the curve in only one other point. It appears thus that there is through any point of the curve a line, called the *tangent*, meeting the curve in two coincident points at this point; and a plane through the tangent, called the *osculating plane*, which meets the curve in three coincident points at this point. The two arbitrary points A, D of the curve being taken, the points B and C are the intersections of the tangents at these points, respectively, with the line in which the osculating planes of the curve, at these points, intersect.

More generally, given any three lines, the equations of three corresponding planes of axial pencils, which are related to one another, having these lines as axes, can be supposed to be of the forms $P-\theta Q=0$, $P_1-\theta Q_1=0$, $P_2-\theta Q_2=0$, where each of P, Q, etc. is a linear function of coordinates x, y, z, t, with any fundamental points of reference. Supposing that the axes, and the manner of relation of the axial pencils, are so general that these three planes do not meet in a line, the point of intersection of these planes has coordinates, found by solving the three linear equations, each proportional to an expression of the form $a\theta^3+b\theta^2+c\theta+d$. We may then suppose x, y, z, t respectively equal to $f_1(\theta), f_2(\theta), f_3(\theta), f_4(\theta)$, where each of these functions is such an expression. There can be no identical relation, $\Sigma A_r f_r(\theta)=0$, where the four coefficients A_r are independent of θ, or the locus of (x, y, z, t) would lie in a plane, which we suppose not to be the case. Wherefore, by reverting the equations, we can express each of θ^3, θ^2, θ, 1 as a linear function of x, y, z, t, say $X=\theta^3$, $Y=\theta^2$, $Z=\theta$, $T=1$. We then have the expression of the curve found above, with X, Y, Z, T as coordinates.

Taking any two chords, l, m, of the cubic, a line, p, can be drawn, from any point, P, of the curve to meet both l and m. This line lies in both the planes $l(P)$, $m(P)$. When P varies on the curve, these planes, by what we have seen, describe related axial pencils, so that the line p describes a quadric surface containing both l and m. Thus a quadric surface can be drawn containing the curve and any two of its chords. The generators of this surface which are of the same system as l and m all meet the generator, p, through any particular point of the curve; the plane containing p and any one of these generators, say n, meets the curve in three points; as the curve lies on the quadric surface, and the plane meets this only in the lines p, n, these three points must lie on these lines; by the construction,

p meets the curve in only one point, P. Wherefore the generators, n, of the same system as l and m, which meet p, are all chords of the curve. We have seen that we may consider related ranges of points upon the curve; we may thus consider involutions of pairs of points upon the curve. In fact the chords, n, meeting the line p, intersect the curve in pairs of points of such an involution, that, namely, determined by the two pairs of points of the curve which are on the chords l and m, respectively. This is clear by considering the cone which projects the curve from P; the two generators of this cone which join P to the two points where one of the chords n meets the curve, lie on a plane passing through the fixed line p; as for conics, this is the necessary and sufficient condition for an involution of pairs of generators of the cone. We thus see that, if we consider any involution of pairs of points of the cubic curve, the chords of the curve, each containing a pair of the involution, are the generators, of one system, of a quadric surface containing the curve. Also, if an arbitrary line, p, be drawn through a point, P, of the curve, not meeting the curve again, any variable plane through p meets the curve in two points, each determining the other, which are a pair of an involution upon the curve. Of the chords joining the pairs of this involution there is one through any point, say O, of the line p. Thus, through any point O of space can be drawn a chord of the curve; for take any point, P, of the curve, and consider, in what has just been said, the line p as being the line OP. Two such chords are not possible through O, since the plane containing these would meet the curve in four points. It appears thus that through any line, p, which meets the curve in one point, and the curve, a quadric surface can be drawn; and further that if a series of such lines p be taken, all containing a point O, these quadric surfaces have all a common generator. More generally it appears that a quadric surface can be drawn to contain the curve and two arbitrary points not lying on the curve; for, taking the chords of the curve passing, respectively, through these points, this is the surface containing the curve, and these two chords.

This last result is in accordance with the fact, which we can reach without difficulty, that the equation of the general quadric surface containing the curve is of the form $\lambda U + \mu V + \nu W = 0$, where U, V, W are definite quadric functions, and λ, μ, ν are arbitrary constants. This is the equation of a general quadric conditioned to pass through seven given points of independent position; for the unconditioned equation of a quadric contains ten terms. In fact, a cubic curve which does not lie wholly upon a quadric surface meets this in six points, as we see by substituting in the equation of the quadric respectively θ^3, θ^2, θ, 1 for the coordinates. Wherefore, a quadric surface prescribed to pass through seven general points of a cubic curve contains the curve entirely.

We may remark at this point: (a), that a cubic curve can be made to pass through six arbitrary points of space, say A, B, C, D, O, Q; for the five points A, B, C, D, O lie on a definite quadric cone of vertex Q, and the five points A, B, C, D, Q lie on a definite quadric cone of vertex O. The curve is the intersection, other than the common generator OQ, of these two cones; (b), that through five general points, say A, B, C, D, E, of a quadric surface, two cubic curves can be drawn which lie entirely upon the surface. For, if l, m be the generators of the surface at E, a definite quadric cone can be constructed, with vertex at E, to contain one of these generators, say l, and also the four points A, B, C, D; the points common to this cone and the quadric, other than the points of the line l, constitute a cubic curve, passing through E. For, let P be any point of the curve of intersection, and p the line drawn from P to meet l and to meet any other fixed generator, k, of the quadric, of the same system as l, so that p is also a generator of the quadric; the quadric is the locus of the points of the lines p, as P varies, and p is the intersection of corresponding planes of two related axial pencils whose axes are l and k, two of these planes being $l(P)$ and $k(P)$. The quadric cone drawn with vertex at E is however the locus of the lines of intersection of corresponding planes of two related axial pencils whose axes are l and (for example) the line EA, two of these planes being $l(P)$ and EAP. By taking k to pass through A, the point P, which is the intersection of the corresponding planes $l(P)$, $k(P)$, EAP, appears as lying on another quadric cone, of vertex A, which is the locus of the intersection of the corresponding planes, $k(P)$, AEP, of two related axial pencils. Another cubic curve lying on the given quadric surface, and passing through the points E, A, B, C, D, is that lying on the cone of vertex E described to contain A, B, C, D and the second generator, m, at the point, E. As we see by considering the intersection of the cone of vertex E, and of the quadric, with an arbitrary plane drawn through l, the former cubic curve has l for a chord; the latter has m for a chord. We see, further: (c), that every cubic curve on a quadric surface has all the generators of one system as chords, meeting every generator of the other system each in one point. For the plane of two intersecting generators of the quadric meets the quadric only on these lines, and meets, therefore, a cubic curve lying on the surface only on these lines. The curve cannot meet either of these lines in three points, since else an arbitrary plane through this line would have no further intersection with the curve, which, therefore, would degenerate into the line itself (taken three times). If the curve meets one of these lines in two points it must meet every generator of the surface, which meets this, in one point; if it meets one of these lines in one point, it must meet every generator of the surface, which meets this, in two points. Thus also, a cubic curve, in (*b*), through E, A, B, C, D, must have either l or m for a chord; there is then no other cubic through these points than those two found in (*b*).

We have defined a cubic curve as the locus of the points common to two quadric cones which have a common generator, and have shewn that quadric surfaces can be drawn through the curve; two of these with a chord of the curve as common generator intersect then further in the curve. Conversely, if we take any two quadric surfaces with a common generator, their remaining intersection is a curve meeting an arbitrary plane in three points. That this is not a more general cubic curve than that above found, may be shewn in detail by proving, as in the above argument under (*b*), that it can be regarded as the intersection of two quadric cones with a common generator.

The construct which is dual to a cubic curve will be that of the aggregate of planes touching two quadric surfaces which have a common generator, other than those passing through this line; or, as follows by the dual of the reasoning above given, of the aggregate of planes touching two conics in different planes which both touch the line of intersection of these planes; or, also, of the aggregate of planes meeting three lines in the corresponding points of three related ranges upon these lines, the ranges being such that the three corresponding points do not always lie in line. There is, evidently, an ∞^1 of such planes, that is, the planes are each determined by a value of a single varying parameter; they are therefore said to constitute a *developable*. And, of such planes, three pass through an arbitrary point, as in the dual case. The aggregate is thus called a cubic developable. Consider this developable as defined by the common tangent planes of two conics in different planes, touching the line of intersection of their planes respectively in A and B. Let P, Q be points, of these two conics respectively, which are such that the tangent lines of these conics at P and Q intersect one another, in the point C, of the line AB. From any point, O, of the line PQ, the conics are projected into two quadric cones; these touch one another along the line OPQ. These cones have four common tangent planes, of which OAB is one, and $OPQC$ is another; this last is to be regarded as two coincident planes; and there is another plane tangent to both cones passing through O; let this plane touch the conics, respectively, in P' and Q', and meet the line AB in C', so that $C'P'$ and $C'Q'$ touch the conics respectively at P' and Q'; let $C'P'$, $C'Q'$ meet CP, CQ, respectively, in M and N, so that MN passes through O, as being on the plane $C'P'Q'$ and intersecting PQ. Then, as O varies on the line PQ, the point C' will vary on AB, and M, N will vary on CP, CQ, respectively. But, as $C'P'M$ is a tangent of one conic, of which AB and

CP are fixed tangents, the ranges (C'), (M) will be related; likewise the ranges (C'), (N) are related. Thus MN envelopes a conic in the plane CPQ, of which CP, CQ are tangents, as also is PQ, since when C' is at C, the points P' and Q' are respectively at P and Q. Let T be the point of contact of this conic with PQ. Then, when O is at T, this third plane of the developable drawn from O, which in general is $OC'P'Q'$, coincides with the former one $OCPQ$. Thus, the developable is such that in any plane of it, (CPQ), there is a line, (PQ), such that, of the three planes of the developable drawn from any point of the line, two coincide with this plane; and on this line there is a point, (T), with the property that the three planes of the developable drawn from it all coincide with this plane. The line is called a *generator* of the developable; it is the dual of the tangent line of the cubic curve above explained. The point T is similarly the dual of the osculating plane of the cubic curve.

It can be proved, however, that, like the conic, the cubic curve is a self-dual construct, the osculating planes forming a cubic developable; and, similarly, the cubic developable is self-dual, the points T above obtained describing a cubic curve, the osculating plane of which, at the point T, is the plane of the developable in which T has been constructed. Thus, not only does the osculating plane of the cubic curve, at a point P, have all its three intersections with the curve coincident at P, but the three osculating planes which can be drawn of the cubic curve through a general point, all coincide, when this point is at P, with the osculating plane at P. The tangent lines of the cubic curve are the generators of the cubic developable which is the aggregate of its osculating planes, and, dually, the tangent line of the locus of the points T is the generating line PQ. The locus of the points T, regarded as arising from the developable, is often called the *cuspidal edge* of the developable.

The proof that the cubic curve is a self-dual construct in the sense explained is very easy with the symbols. The curve being regarded as the locus of the point $(\theta^3, \theta^2, \theta, 1)$, the plane whose equation is $x-py+qz-rt=0$, where $p=\alpha+\beta+\gamma$, $q=\beta\gamma+\gamma\alpha+\alpha\beta$, $r=\alpha\beta\gamma$, evidently contains the three points of the curve for which $\theta=\alpha$, $\theta=\beta$, $\theta=\gamma$. Thus the plane whose equation is

$$x-3\theta y+3\theta^2 z-\theta^3 t=0$$

is the osculating plane at the point θ. Evidently three such planes pass through an arbitrary point. Further, these planes meet the plane $t=0$, the osculating plane at $\theta^{-1}=0$, in the lines

$$x-3\theta y+3\theta^2 z=0,$$

which are tangent lines of the conic $t=0$, $3y^2=4xz$, the point of contact of one such line with this conic being $(3\theta^2, 2\theta, 1, 0)$; and these planes meet the plane $x=0$, the osculating plane at $\theta=0$, in the lines $3y - 3\theta z + \theta^2 t = 0$, which are tangent lines of the conic $x = 0$, $3z^2 = 4yt$, the point of contact of one such line with this conic being $(0, \theta^2, 2\theta, 3)$. The generator of the developable formed by these planes, being the line joining the points of contact of the variable plane with these conics, is the line whose general point is expressed by $(3\theta^3, 2\theta^2 + \lambda\theta^2, \theta + 2\lambda\theta, 3\lambda)$; this is immediately proved to be the tangent of the original curve at the point θ. The general point of a tangent line of the cubic, of which, as has been remarked, the coordinates are $3\theta^3, (\lambda + 2)\theta^2, (2\lambda + 1)\theta, 3\lambda$, is easily shewn to describe a surface of which the equation is $V^2 = 4UW$, where $U = xz - y^2$, $V = xt - yz$, $W = yt - z^2$. Thus four tangent lines of the cubic curve meet an arbitrary line; or, through an arbitrary line can be drawn four planes each of which contains a tangent line of the cubic curve. Dually, the generators of the cubic developable meet an arbitrary plane in the points of a curve of which four points lie on an arbitrary line of this plane. This plane meets the cuspidal edge, which is a cubic curve, in three points. Each of these three points is a *cusp* of this plane quartic curve; that is, it is a double point of this curve, or a point such that every line, in the plane of the curve, drawn through this point, meets the curve in two coincident points at this point (and, therefore, meets the quartic curve in two other points); and it is a double point at which the two tangents of the curve coincide. It is for this reason that the name cuspidal edge is given.

As the notion of a developable, and its cuspidal edge, is fundamental for the understanding of the dual of a curve, it may be desirable to refer to a concrete representation. Let two pieces of paper, laid one over the other, be simultaneously cut by a pair of scissors along an open portion of an arbitrary curve, say an arc of a circle, which thus becomes a portion of the boundary of each piece of paper. Let the two pieces of paper be now connected along this arc by means of tissue paper gummed to both pieces of paper, to the upper side of the upper piece of paper and to the lower side of the lower piece. Now let tangent lines be drawn to this arc on the upper side of the upper piece of paper, and on the lower side of the lower piece; let the portion of a tangent line on the upper piece which lies on one side of the point of contact with the arc be coloured red, and the continuation of this tangent which lies on the lower piece be also coloured red; let the other portion of this tangent line on the upper piece be coloured blue, and the continuation of this tangent which lies on the lower piece be also coloured blue; let this be done, in a continuous way, for a succession of tangent lines. If now the two pieces of paper be lifted by a corner of the upper piece, not lying on the arc, the paper being stiff enough for the red tangent lines, which are partly on one sheet and partly on the other, to remain straight, the blue lines being allowed to curve, and the arc not remaining in a plane, these red lines will represent generators of a developable surface of which the arc is a portion of the cuspidal edge; at each point of this edge there will be a red tangent line. (Thomson and Tait, *Natural Philosophy*, Part I (1879), p. 114, § 149.)

We may usefully consider the properties of a cubic developable in more detail, taking the steps dual to those followed for the cubic curve. As a line joining two points of the curve is called a chord, let the line of intersection of two planes of the developable be called an *axis*. As for the cubic curve we took two quadric cones with a common generator, so now we take two conics, say Σ and Σ', in different planes, say ϖ and ϖ', both conics touching the line of intersection of their planes. A plane of the developable will be any plane touching both these conics: it will meet the two planes in tangents of the two conics, of which each will determine the other by the fact that they intersect on the line of intersection of the two planes. In particular each of the two given planes will be a plane of the developable. In the case of the cubic curve two chords determine a quadric surface containing the curve, and from any point of the cubic there passes a line, intersecting these two chords and all the other chords which are generators of this quadric, of the same system; these chords determine an involution of pairs of points of the cubic curve, and project from any point of the curve into chords of a conic meeting in a point. Conversely, if an arbitrary line be drawn from any point of the cubic curve, and the chord be drawn to the cubic from any point of this line, these chords are generators of a quadric surface. In the dual figure, if any point, P, be taken in the plane ϖ, two tangents can be drawn from P to the conic Σ, meeting the line of intersection of the planes ϖ, ϖ' in two points; from each of these can be drawn, beside this line of intersection, a tangent line to the conic Σ', and these meet in a point, P', of the plane ϖ'. The line PP' is an axis of the developable, and the construction establishes a $(1, 1)$ correspondence between the points of the planes ϖ, ϖ', which is in fact of the kind considered (Vol. I, p. 148). In the original figure there is similarly a correspondence between the planes drawn through the vertex of one of the two cones with the planes drawn through the vertex of the other cone. Consider now various points P lying on an arbitrary line of the plane ϖ. These will be the dual of various planes passing through the vertex of one of the two cones of the original figure, all having in common a line through this vertex. The axes PP' will then be all generators of a quadric surface, touching all the planes of the developable; the pairs of these planes through the various axes PP' will be pairs of an involution of planes of the developable; the locus of the points P' will be a line, which, like the line on which P is taken, is a generator of the quadric surface generated by the lines PP'. So long as this quadric surface does not degenerate, these two corresponding lines of the planes ϖ, ϖ' do not intersect. This happens, however, when the line on which P varies is a tangent line of the conic Σ; then the locus of P' is a tangent of the conic Σ', and the plane of these intersecting lines is a plane of the developable. In the original figure the corresponding statement is: if from two arbitrary points as centres lines be drawn generating two related central systems, so that the points in which these lines meet a plane, not passing through the centres, are two related plane systems, then corresponding lines, one from each centre, do not in general intersect; when, however, they do, their point of intersection lies on a cubic curve, and the intersecting lines are generators of two quadric cones which intersect in this curve. This gives a new definition of the curve. The theorem relates to the most general way of relating two central systems of lines. What particular cases can arise will be seen from the detailed consideration of two related plane systems given later.

Examples in regard to the cubic curve in space. *Ex.* 1. The tangent line of the curve at any point is a generating line of the quadric cone projecting the curve from this point; the osculating plane of the curve at this point is the tangent plane of this cone along this generator.

Ex. 2. The osculating planes at any three points of the curve meet in a point lying in the plane containing these points. If the point of meeting be (x_0, y_0, z_0, t_0), the plane has the equation $xt_0 - 3yz_0 + 3zy_0 - tx_0 = 0$. This is the focal plane of the point (x_0, y_0, z_0, t_0) in regard to the linear complex expressed by $l + 3l' = 0$; in particular when (x_0, y_0, z_0, t_0) is on the curve, it is the osculating plane of the curve at this point, all the tangent lines of the curve belonging to this linear complex. Further, if any line be taken, and the four tangents of the cubic curve which meet this line, the second of the two transversals of these four tangents is the polar line of the first in regard to this linear complex.

Ex. 3. Since the four planes which join a chord of the cubic curve to four arbitrarily chosen fixed points, A, B, C, D, of the curve form a pencil related to that of the planes joining these points to any other chord, it follows that all the chords of the curve belong to a quadratic tetrahedral complex depending upon A, B, C, D (see Chap. II, p. 99 above). If the values of the parameter θ which belong to the points A, B, C, D be, respectively, a, b, c, d, the coordinates, relative to A, B, C, D, of any point of the cubic, can be taken to be $(\theta - a)^{-1}, (\theta - b)^{-1}, (\theta - c)^{-1}, (\theta - d)^{-1}$, and the line coordinates l, l', etc., of any chord are then such that

$$ll'/(b-c)(a-d) = mm'/(c-a)(b-d) = \text{etc.}$$

Ex. 4. Shew that the general equation of a quadric surface containing the cubic curve, of which the general point is $(\theta^3, \theta^2, \theta, 1)$, is $A\xi + B\eta + C\zeta = 0$, where $\xi = xz - y^2$, $\eta = xt - yz$, $\zeta = yt - z^2$, one system of generators of this meeting the curve in the pairs of points (θ), (ϕ), of the involution given by $A\theta\phi + B(\theta + \phi) + C = 0$. Further, that the line coordinates of the chord of the curve passing through any point (x, y, z, t) are, respectively, $\eta^2 - \xi\zeta$, $\eta\zeta$, ζ^2, $-\xi\zeta$, $\xi\eta$, $-\xi^2$.

Ex. 5. We have seen that a cubic curve lying on a quadric surface meets every generator of the surface of one system in one point, and every generator of the other system in two points. Thus, taking the intersections, with a fixed arbitrary plane section of the quadric, of the generators through the points of the curve, there is determined on this section a (1, 2) correspondence (see Vol. II, p. 134). Conversely, such a correspondence upon any plane section of the quadric determines two cubic curves lying on the surface.

Ex. 6. If A, B, C, A', B', C' be any six points of the cubic curve, the three quadric surfaces containing the curve, of which the first contains the two chords BC', $B'C$, the second contains the two chords CA', $C'A$, and the third the two chords AB', $A'B$, have all in common, as generator, another chord of the curve (cf. Vol. II, p. 16). Further, any chord of the curve is met in pairs of points of an involution by the quadrics through the six points.

Ex. 7. A cubic developable is formed, we have seen, by the planes joining corresponding points of three related ranges of points, on three lines, of which no three corresponding points are in line. If we have five lines l, m, l', m', p, in space, of which no two intersect, these lines not having a common transversal, and, from a variable point of p, the transversal be drawn to l and m, and also the transversal to l' and m', the plane of these two transversals describes a cubic developable, of which the five given lines are axes (cf. Ex. 22, below). If P, Q describe related ranges upon a conic, and R describe a range related to both, on a line not in the plane of the conic, prove that, in general, the planes PQR form a cubic developable. Also, if P, Q, R describe three related ranges of points on a cubic curve, the planes PQR form a cubic developable. Further, if the points P, Q of a cubic curve are in (2, 2) correspondence, so that to any position of P corresponds two positions, Q and Q', of Q (and conversely), then the planes PQQ' form a cubic developable, and the point of this plane which is the intersection of the osculating planes at P, Q, Q', of the original cubic, describes another cubic curve.

Ex. 8. From the variable point P, of a fixed straight line, the chord PQR is drawn to a cubic curve, and thereon is taken the point, T, which is the harmonic conjugate of P in regard to R and Q. Thus T is conjugate to P in regard to all quadrics containing the curve. Prove that the locus of T is a cubic curve, meeting the given cubic in four points. Further, if $P_0Q_0R_0T_0$ be a particular position of the line $PQRT$, and O be a fixed point of the cubic, and the transversal from O to $PQRT$ and $P_0Q_0R_0T_0$ meet the latter in

P', prove that the ranges (P), (P') are related, the planes joining O to the chords $PQRT$ being tangents of a quadric cone. This cone also touches the fixed line, on which P is taken. Also, for a fixed position of P, the polar planes of P in regard to the cones which project the cubic from the points of the curve, envelope a quadric cone whose vertex is T.

Ex. 9. If six lines, of which no two intersect, be all met by another line, shew that a variable plane, whose points of intersection with the seven lines lie on a conic, describes a cubic developable, of which the first six lines are axes. (Cf. Ex. 7, p. 195, below.)

Ex. 10. It is a familiar fact that if $f, = ax^2 + 2hxy + by^2$, and $f_1, = a_1x^2 +$ etc., be two quadratic forms in x, y, and we take the two values of x/y for which, for proper values of λ/λ_1, the form $\lambda f + \lambda_1 f_1$ is a perfect square, the quadratic $F, = Ax^2 +$ etc., which vanishes for these values of x/y, is such that both the expressions $Ab - 2Hh + Ba$, $Ab_1 - 2Hh_1 + Ba_1$ vanish (Vol. II, p. 8). More generally, if $f_r, = a_{r0}x^n + na_{r1}x^{n-1}y + \ldots + a_{rn}y^n$, for $r = 1, 2, \ldots, n$, be n forms each homogeneous of the nth order in x, y, there are n sets of values of the ratios of $\lambda_1, \lambda_2, \ldots, \lambda_n$ such that the form $\lambda_1 f_1 + \ldots + \lambda_n f_n$ is the nth power of a linear form, $px + qy$. The form F, of the nth order, which vanishes for the n values of x/y for which these linear forms vanish, is expressed by a determinant of $(n+1)$ rows and columns of which the $(s+1)$th row consists of the elements $a_{1s}, a_{2s}, \ldots, a_{ns}, y^{n-s}(-x)^s$. Denoting this form by $A_0x^n + nA_1x^{n-1}y +$ etc., each of the n expressions

$$A_0 a_{r,n} - nA_1 a_{r,n-1} + \binom{n}{2} A_2 a_{r,n-2} - \text{etc.}$$

is zero, and F is the form which is *apolar* to all the forms $f_1, \ldots, f_n$ (cf. Vol. II, p. 114, etc.).

Ex. 11. Any plane drawn through an arbitrary point, O, has an equation of the form $\lambda_1 U_1 + \lambda_2 U_2 + \lambda_3 U_3 = 0$, where

$$U_r = a_r x + 3b_r y + 3c_r z + d_r t \quad (r = 1, 2, 3),$$

and $U_r = 0$ are three planes of general position through this point. If f_r denote what U_r becomes by putting $\theta^3, \theta^2, \theta, 1$, respectively, for x, y, z, t, it is in accordance with Ex. 10 that three osculating planes of the cubic curve can be drawn through the point O. By that example, the three points of the cubic curve, say A, B, C, where these osculate the curve, correspond to a cubic form which is apolar to that corresponding to the three intersections of the curve with any plane drawn through O. We call O the pole of the plane ABC. Thus, in particular, any plane drawn through the chord of the cubic which passes through O, meets the curve in three points giving rise to a cubic which is apolar with the cubic associated with the points

A, B, C. Thus we infer that the Hessian of this last cubic (A, B, C), corresponds to the ends of the chord of the curve drawn through O (cf. Vol. II, p. 124). Further, it follows, as on p. 136 of Vol. II, that an involution of sets of three points upon the cubic curve is given by the intersections of the curve with planes through an arbitrary fixed line; and, then, as here, that if one of these sets of points, A, B, C, be such that the pole, O, of the plane ABC, lies upon this line, then this is the case for every one of the sets, every two of which are then apolar. In general, the poles of planes passing through a line (l, m, n, l', m', n') lie on the line $(-3l', m, n, -\frac{1}{3}l, m', n')$.

Ex. 12. We have seen in Vol. II (p. 124) that the Hessian points of three points, A, B, C, of a conic, may be obtained by taking the three points such as the intersection of the line BC with the tangent of the conic at A, and then the points where the line containing these three points meets the conic. When the cubic curve is projected to a conic, on an arbitrary plane, from any point, D, of itself, any point of this conic may evidently be represented by the parameter, θ, appropriate to the point of the cubic of which it is the projection. From this we have a construction for the Hessian pair of the three points A, B, C of the cubic, and for the chord of this curve drawn through the pole, O, of the plane ABC: namely, if we take the line in which the plane DBC is met by the plane joining D to the tangent line of the curve at A, the three such lines lie on a plane through D, whose intersections, other than D, with the curve, are the Hessian points of A, B, C. The chord joining these points meets the plane ABC in its pole, O. Or, if the tangent lines of the curve at A, B, C, respectively, meet the planes DBC, DCA, DAB in A', B', C', the plane $A'B'C'$ contains D, and contains the Hessian chord of A, B, C.

Ex. 13. Now let, again, A, B, C, D be any four points of the cubic curve. As the osculating planes of the curve at A, B, C meet in a point, O, of the plane ABC, and similarly for the osculating planes of any three of these four points, the four points such as O form a Moebius tetrad both in- and circumscribed to A, B, C, D (Vol. I, p. 61). But, also, if A', B', C', D' be the points where the tangent lines of the curve at the points A, B, C, D, respectively, meet the planes DBC, etc., it follows from the preceding example: (i), that the points A', B', C', D' form a tetrad in- and circumscribed to A, B, C, D, the plane $A'B'C'$ containing D, etc., and, (ii), that the points A', B', C', D' are a tetrad which is also in- and circumscribed to the tetrad of the four poles such as O, the plane $A'B'C'$ containing O, while D' lies in the osculating plane at D. Thus, if L be the pole of the plane DBC, namely the intersection of the osculating planes at D, B, C and, similarly, M, N be the poles of the planes DCA, DAB, respectively, the three tetrads A, B, C, D; L, M, N, O;

A', B', C', D' are such that every two are a pair of Moebius tetrads. It can be shewn to follow from this that the four planes $AA'L$, $BB'M$, $CC'N$, $DD'O$ meet, in threes, in a tetrad in- and circumscribed to each of the three tetrads referred to. For this it is sufficient to shew that the three planes $AA'L$, $BB'M$, $CC'N$ meet in a point lying on the planes ABC, $A'B'C'$ and in the osculating plane at D. The theorem, however, is a case of that enunciated in the succeeding example.

Ex. 14. The general tetrad of points, A_1, B_1, C_1, D_1, forming a Moebius tetrad with four points A, B, C, D, are a set whose coordinates, relative to A, B, C, D, are capable of the forms

$$(0, -n_1, m_1, -l_1), (n_1, 0, -l_1, -m_1), (-m_1, l_1, 0, -n_1), (l_1, m_1, n_1, 0),$$

depending on the ratios of the arbitrary quantities l_1, m_1, n_1. Another such tetrad, depending on l_2, m_2, n_2, form a Moebius tetrad with A_1, B_1, C_1, D_1 if $l_1 l_2 + m_1 m_2 + n_1 n_2 = 0$. Thus a set of four tetrads of which every two are Moebius tetrads arises if $(l_1, m_1, n_1, 0)$, $(l_2, m_2, n_2, 0)$, $(l_3, m_3, n_3, 0)$ are a self-polar triad of points in regard to the conic $t = 0$, $x^2 + y^2 + z^2 = 0$.

Two tetrads, A_1, B_1, C_1, D_1 and A_2, B_2, C_2, D_2, each both inscribed and circumscribed to A, B, C, D, can also be obtained by taking AA_1, BB_1, CC_1, DD_1 as the transversals to two arbitrary generators, of the same system, of an arbitrary quadric, in regard to which A, B, C, D are self-polar, and AA_2, BB_2, CC_2, DD_2, as the transversals to two other generators of the same system. When the two latter generators are harmonic conjugates in regard to the two former, then the two tetrads, A_1, B_1, C_1, D_1 and A_2, B_2, C_2, D_2, are also mutually inscribed. In this case the four planes such as DD_1D_2, AA_1A_2 give rise, by their intersections, to a fourth tetrad inscribed and circumscribed to the previous three tetrads. (See pp. 67, 68, above.)

Ex. 15. Four such tetrads are also representable, for any values of a, b, c, d, by the rows (or by the columns) of the scheme

$$\begin{matrix} iPi & iPj & iPk & iP \\ jPi & jPj & jPk & jP \\ kPi & kPj & kPk & kP \\ Pi & Pj & Pk & P \end{matrix}$$

where $P = ai + bj + ck + d$, the symbols i, j, k are such that

$$i^2 = j^2 = k^2 = -1, \quad jk = -kj = i, \quad ki = -ik = j, \quad ij = -ji = k,$$

and $xi + yj + zk + t$ represents the point (x, y, z, t).

Ex. 16. We may consider a general involution of sets of four points upon the cubic curve, given (Vol. II, p. 136) by the condition $U + \lambda V = 0$, for varying values of λ, where U, V are each polynomials of the fourth order in the parameter θ which gives a point of the curve. It can be shewn: (a), that the tetrads of points of the cubic so obtained are all self-polar in regard to a certain quadric; (*b*), that the planes joining three points of the tetrad are those of a cubic developable. In fact, if θ_1, θ_2 be two roots of the quartic equation, so that $U_1+\lambda V_1=0$, $U_2+\lambda V_2=0$, the identity $(U_1V_2-U_2V_1)/(\theta_1-\theta_2)=0$, which does not depend on λ, is linear in regard to both the two sets (x_1, y_1, z_1, t_1), (x_2, y_2, z_2, t_2), where $x_1=\theta_1^3$, $y_1=\theta_1^2$, $z_1=\theta_1$, $t_1=1$, $x_2=\theta_2^3$, etc., and is symmetrical in regard to the two points with these coordinates. Thus it expresses that these points are conjugate in regard to a definite quadric. Further, denoting this relation by $(\theta_1, \theta_2)=0$, we also have $(\theta_1, \theta_3)=0$, $(\theta_1, \theta_4)=0$. Thus the coordinates of the plane joining the points $(\theta_2, \theta_3, \theta_4)$, which are $1, -(\theta_2+\theta_3+\theta_4), \theta_2\theta_3+\text{etc.}, -\theta_2\theta_3\theta_4$, appear as proportional to four cubic polynomials in θ_1. Therefore, as θ_1 varies, the plane $(\theta_2, \theta_3, \theta_4)$ describes a cubic developable. This may be spoken of as the polar reciprocal of the given cubic curve in regard to the quadric surface obtained. The theorems, however, arise incidentally below (Ex. 28).

Ex. 17. If a cubic curve in space, and a conic, have three points in common, a quadric surface exists containing both curves. Also, if two cubic curves in space have five points in common, a quadric surface exists containing both curves; this is not necessarily the case for two cubic curves with only four common points. Prove, further, that two cubic curves lying on a quadric surface have four common points, or five common points; the former is the case when the curves are such as to have the lines of the same system of generators of the quadric as chords, the latter when this is not so. Shew that in the latter case the cubic curves have no common chord other than a join of two of the five common points.

Ex. 18. It has been shewn that a cubic curve can be described to contain six arbitrary points. Shew that a cubic curve can be constructed to pass through five arbitrary points and have a given line as chord. In fact, two independent quadrics can be constructed to pass through the five points and have the line as a generator. Or, if coordinates be taken with four of the five given points as points of reference, a transformation of the form

$$x'=Ax^{-1},\ y'=By^{-1},\ z'=Cz^{-1},\ t'=Dt^{-1},$$

changes the given line into a cubic curve through the four points of reference; from the point which arises, by this transformation, from the fifth given point, a single chord can be drawn to this curve. The curve which is sought is that obtained from this chord by applying the inverse of the transformation referred to.

Ex. 19. Now consider the problem of describing a cubic curve through four given points to have two given lines as chords. Through such a curve would pass a quadric surface containing the lines as generators; in general, however, a quadric surface cannot be drawn through four given points to contain also two given lines. The construction of the cubic curve is, therefore, impossible in general. Suppose however the given lines are two generators, of the same system, of a quadric surface containing the four given points. Then we have seen that a cubic curve can be drawn to satisfy the given conditions, lying on the quadric surface and passing through a fifth arbitrary point of this surface (p. 129, above). Consider the case when the given lines intersect.

Ex. 20. Next suppose we are given three points and three lines of general position. We can clearly describe a cubic curve to have these lines as chords, containing the three points. We have only to consider related axial pencils of planes having these lines as axes, these pencils being related by the fact that three corresponding planes meet in each of the three given points.

Ex. 21. If we are given two points and four lines, of which no two intersect, a cubic curve can be constructed passing through the two given points to have each of the four given lines as a chord. For, let A, B be the given points and l, m, l', m' the given lines. Consider the transversal, a, drawn from A to meet l and m, and the transversal, a', drawn also from A, to meet l' and m', and the plane of these two transversals. Consider also the transversal, b, drawn from B to meet l and m, and the transversal, b', also drawn from B, to meet l' and m', and the plane of these two transversals. These two planes meet in a line, n, which is the only line other than AB which meets the four lines a, a', b, b'. If a cubic curve exists through A, B with l, m, l', m' as chords, it lies on a quadric containing l and m, of which therefore a, b are also generators; for each of a, b contains, beside, the point A or B, two other points of the quadric, lying, respectively, on l and m. The curve equally lies on a quadric containing l' and m', of which therefore a', b' are also generators. These two quadrics, having this cubic in common, must further intersect in a line, which then is a chord of the curve, and of the same system of generators of the first quadric as are l, m; it will therefore meet a and b. Similarly it will meet a', b'. The common generator is, therefore, the line n. Conversely, let a quadric, which is unique, be constructed to contain the lines l, m, n; this quadric then contains a, b and, therefore, A and B. Similarly, the quadric containing the lines l', m', n contains a', b' and, therefore, A and B. These two quadrics, having n in common, determine the cubic curve required, as their remaining intersection.

Ex. 22. If we are given one point, and five lines of which no two intersect, these lines not having a common transversal, a cubic curve exists passing through the given point of which the five given lines are chords. Denote the point by A, and the lines by l, m, l', m', p. Remark first that, by means of the two given pairs of lines, l, m and l', m', a definite plane is determined through any point, P, of space, namely the plane of the transversals drawn from P to l, m and from P to l', m'. Denote this plane by (P). As P varies on the line, p, this plane, describing related ranges on l, m, l', m', p, describes a cubic developable, of which these lines are axes. Of this developable there is then one definite axis on the plane (A), say the line n; this is the intersection of two planes of the developable, passing through positions of P, say B and C, upon the line p; the three planes of the developable which pass through B consist of the plane (B), and the two planes intersecting in the axis p. Evidently, now, A is the intersection of a line meeting l, m, n and a line meeting l', m', n; as, similarly, are B and C. The two quadric surfaces, with n as common generator, determined, respectively, by l, m, n and l', m', n, thus meet in a cubic curve having l, m, l', m' as chords, passing through A, B, C; this curve, therefore, has also p as a chord, and is the curve desired.

It may be remarked, in anticipation, that when the five given lines have a common transversal (in which case there is a position of P on p for which the plane (P) is indeterminate) there are cubic curves in infinite number having these lines as chords; but these all lie on a surface (a cubic surface; *v.* p. 175); one such cubic curve can be drawn passing through two points of this surface.

Ex. 23. If we have two cubic curves in space, of general position, there are ten lines each of which is a chord of both curves. To prove this, first notice that through any point, O, of one cubic, ρ, can be drawn six chords of this curve each of which meets the other curve, σ. For the points of the curve, into which the curve σ is projected from O on to an arbitrary plane, have coordinates proportional to cubic polynomials of the parameter which gives the position of a point of the curve σ; if these be substituted in the equation of the conic into which the curve ρ is projected from O, the result is an equation of order six. Now take any variable point, P, of the curve ρ, and draw from it the chord to the curve σ, meeting it in Q and Q'. From Q draw the chord to the curve ρ, and so from Q'. The point P of the curve ρ thus gives rise to four other points of this curve; let P' be one of these. To pass back from P' to P, we must first draw from P' a chord of ρ which meets σ, say in Q, which is possible in six ways; and then from Q we must draw a chord of σ which meets ρ, in the point P, also possible in six ways. The point P' thus arises from thirty-six different possible positions of P. Thus, if θ, θ' be the parameters giving the positions of P, P', respectively, upon the curve ρ, there is a relation $(\theta, \theta')=0$, an integral polynomial in both θ and θ', of order 36 in

θ and of order 4 in θ'. Or, as it is usual to say, there is a correspondence (36, 4) between P and P' (cf. Vol. II, p. 134). The values of θ and θ' are therefore the same for $36+4$ or 40 values of θ. When this is the case the chord PQQ', drawn from the point P, of ρ, to meet σ in Q and Q', meets ρ again, say in the point H. This common chord however arises in four ways in the construction given: not only when P coincides with a corresponding position P' arising with Q as intermediary, but also when P coincides with a corresponding position arising with Q' as intermediary; and not only when P coincides with P', but also when H coincides with a corresponding position, arising through Q or Q'. The number of common chords of the two curves is therefore one quarter of 40, namely is ten.

Ex. 24. We can hence shew that, if we be given six lines, of general position, there are six cubic curves all having these six lines as chords. Let these lines be l, m, l', m', p, q. We have shewn in Ex. 22 that a cubic curve having the five lines l, m, l', m', p as chords has, as another chord, an axis of the cubic developable described by the planes (P), there explained, as P moves on p. Similarly this cubic, if it have q also as a chord, has as another chord an axis of the cubic developable described by the planes (Q), as Q moves on q. This other chord is the same in both cases, being that meeting the curve in the common pair of the two involutions determined thereon by l, m and by l', m'. These two cubic developables have, however, by the dual of Ex. 23, ten common axes. Of these, four are the lines l, m, l', m'. There remain then six others. Taking any one of these, say n, the cubic required is the locus other than n, of the common points of the quadrics (l, m, n), (l', m', n).

The theorem that there are six cubic curves having six given chords is subject to modification when these chords are not lines of general position. For instance, when the six lines have a common transversal, it may be shewn that there is only one proper, or undegenerate, cubic curve having these as chords. Or again, when every five of the six given lines have a common transversal, there is an infinite number of such curves. Or, as another case, when the six lines, say a, b, c, l, m, n, are such that each of the three sets of five, b, c, l, m, n; c, a, l, m, n; a, b, l, m, n, have a common transversal, there is no proper cubic curve with the six lines as chords; in this case six degenerate curves, each consisting of three lines of which one meets the other two, can be found.

Ex. 25. The theory of the last preceding examples is essentially due to Cremona (*Crelle's J.*, Vol. LX, 1862, p. 191. See also Cayley, *Papers*, Vol. VII, p. 170, 1870). Another method of proof employs a method of transformation which it is desirable to explain. We have already above employed a reversible cubic transformation $x' = Ayzt$, $y' = Bzxt$, $z' = Cxyt$, $t' = Dxyz$, which changes a line into a cubic curve having four fixed points. A reversible cubic transformation of much generality is obtained by taking three fixed quadric surfaces, and making correspond to any point, (x, y, z, t), the point of intersection, (x', y', z', t'), of the polar planes of (x, y, z, t) in regard to these surfaces. Then, as (x, y, z, t) describes any line, the three polar planes describe axial pencils of planes, all related to the range of points (x, y, z, t), whose axes are the polar lines of the line in regard to the three quadric surfaces. Thus (x', y', z', t') describes a cubic curve, having these lines as chords. We can however choose the three quadric surfaces so that the cubic curve so obtained has four fixed lines as chords, which are independent of the line of which the cubic curve is the transformation, and may be taken quite arbitrarily; conversely, in this case, any cubic curve having these four lines as chords is transformed into a line. By the transformation so obtained, the theorem, for example, that a cubic curve can be described to pass through two given points and have four given lines as chords, becomes the theorem that a line can be drawn through two points. Again, the theorem that a cubic curve can be drawn through a given point to have five given lines as chords, when these lines have no common transversal, becomes the theorem that a line can be drawn through an arbitrary point to be a chord of a given cubic curve. Lastly, the cubics having six given lines, of general position, as chords become, by the transformation, the common chords, other than four given ones, of two given cubics.

We may obtain the particular cubic transformation in question, in dual form, in a simple manner, by proceeding as follows: Taking the Moebius tetrads, A, B, C, D and A', B', C', D', the latter points being of equations relative to the former, respectively $U=0$, $V=0$, $W=0$, $P=0$, where $U = c'v - b'w - ap$, $V = -c'u + a'w - bp$, $W = b'u - a'v - cp$, $P = ua + vb + wc$, the point-pairs (A, A'), (B, B'), (C, C'), (D, D') are not independent, their equations being connected by the identity $uU + vV + wW + pP = 0$. Thus the poles of an arbitrary plane in regard to these four point-pairs lie in one plane, as do the poles of an arbitrary plane in regard to three independent quadrics. The pole of a plane in regard to a point-pair, such as A, A', is the point of the line AA' which is the harmonic conjugate, in regard to A and A', of the point where the plane meets the line AA'. The new plane may be taken to correspond to the original plane, which, conversely, corresponds to the new plane by the same rule. It is this correspondence which is effected by the transformation in question. Consider, now, all the planes which correspond, by this rule, to the planes, say α, which can be drawn through an arbitrary fixed line, say l. If a plane α meet the lines AA', BB', CC', respectively, in U, V, W, the ranges (U), (V), (W), being sections of an axial pencil of planes, are related; therefore, the ranges (U'), (V'), (W'), where U', V', W' are the points lying respectively on AA', BB', CC' for which the ranges A, U, A', U'; B, V, B', V'; C, W, C', W' are harmonic, are also related. Whence, the plane $U'V'W'$ describes a cubic developable. Of this developable each of the lines AA', BB', CC', DD' is an axis. For instance, there are two planes α, passing through l, of which the corresponding planes both pass through DD', those, namely, containing l and the two common transversals of the four lines l, AA', BB', CC'; it is, in fact, a property of two Moebius tetrads that, when U, V, W are in line, the plane $U'V'W'$ contains DD' (Chap. I, p. 67, Ex. 20). The two planes α may also be defined as the two tangent planes through l of the quadric described to have the lines AA', BB', CC' as generators.

It is now to be shewn, conversely, that given any four lines, of which no two intersect, it is possible to find two Moebius tetrads, A, B, C, D and A', B', C', D', of which A, A' are on the first line, B and B' on the second, and so on. This is always possible; formulae applicable for this purpose are given in Ex. 22, p. 68; it is only necessary to express that the four pairs of points A, A'; B, B', etc., each harmonic conjugates in regard to the points in which their join is met by the two common transversals of the four joins, are not independent.

Dually, the polar planes of a point, P, in regard to the plane pairs, $x\xi=0$, $y\eta=0$, $z\zeta=0$, $t\tau=0$, where

$$\xi = a't + bz - cy, \ \ldots, \quad \tau = -(a'x + b'y + c'z),$$

meet in a point P', which, as P describes a line, l, of general position, describes a cubic curve having as chords the four lines, k, such as $x=0=\xi$. The points of intersection of the cubic with one of these lines, k, correspond to the points where the original line, l, meets the quadric containing the other three of the lines k.

An incidental consequence of the last remark may be referred to in passing: In order that it may be possible to describe a cubic curve touching each of the four given lines, it is necessary and sufficient that the four quadrics each containing three of these lines should all be touched by a line. It can be shewn that they are then touched by an infinite number of lines, all generators of a quadric, of the same system. There is then an infinite number of cubic curves touching the four lines. This is also clear by considering that if we take four tangents of a cubic curve there is an infinite number of linear complexes to which these four tangents belong; the poles, in regard to any one of these complexes, of the osculating planes of the given cubic curve, describe another cubic curve also touching the four given tangents. The problem of finding a common tangent line of the four quadrics which each contain three of four non-intersecting lines is the same, in fact, as of finding a plane section of a quadric surface, which contains four points A, B, C, D, which shall touch the sections of this surface by the four planes BCD, CAD, ABD, ABC. For the surface

$$fyz + gzx + hxy + t(ux + vy + wz) = 0,$$

the condition that this shall be possible can be shewn to be $l^{\frac{1}{2}} + m^{\frac{1}{2}} + n^{\frac{1}{2}} = 0$, where $l = \sigma(fu)^{\frac{1}{2}}$, $m = \epsilon(gv)^{\frac{1}{2}}$, $n = \sigma\epsilon(hw)^{\frac{1}{2}}$, in which $\sigma^2 = 1$, $\epsilon^2 = 1$. If $(l_1, m_1, n_1, l_1', m_1', n_1')$, etc., be the line coordinates of four tangent lines of a cubic curve, and

$$\varpi_{12} = l_1 l_2' + l_2 l_1' + \text{etc.},$$

it can be proved that $\alpha^{\frac{1}{4}} + \beta^{\frac{1}{4}} + \gamma^{\frac{1}{4}} = 0$, where $\alpha = \varpi_{23}\varpi_{14}$, etc.

Ex. 26. There is another problem of construction for cubic curves which may be referred to: Given four lines of general position, and upon each of them an involution of pairs of points, there are two cubic curves having these four lines as chords whose points of meeting with each of these lines are a pair of the involution lying thereon. This may be deduced from a theorem, given by Chasles (*Liouville's J.*, t. IV, 1839, p. 348), that the lines of space which meet the generators of one system of a quadric surface in points lying on pairs of a given involution, of pairs of these generators, are lines of a linear complex. (Above, Ex. 15, p. 67.)

Ex. 27. Consider a cubic curve, θ, and a quadric surface, Σ. Let l, m be any two chords of the curve. To any plane, α, drawn through l, meeting the curve again in a point P, can be drawn a plane, β, through m, conjugate to the former in regard to Σ. As α varies, describing an axial pencil about l, the plane β will describe a related axial pencil about m, and if β meet the curve again in Q, the ranges (P), (Q) will be related. The pairs of points P, Q will then belong to an involution on the curve if, for one pair, not only Q corresponds to P but P also corresponds to Q (Vol. II, p. 2); that is, if not only the planes lP, mQ be conjugate in regard to Σ, but also the planes lQ, mP be conjugate.

Now suppose that upon the curve θ there are four points, A, B, C, D, forming a self-polar tetrad in regard to Σ. Take any point, T, of the curve, and let its polar plane in regard to Σ meet the curve in X, Y, Z. We can then shew that T, X, Y, Z are a self-polar tetrad in regard to Σ. For consider pairs of planes, conjugate to one another in regard to Σ, drawn respectively through the chords BC and YZ; the plane joining YZ to D, which is the pole of the plane BCA, is conjugate to this plane, joining BC to A; similarly the plane joining YZ to A is conjugate to the plane joining BC to D. Wherefore, these pairs of planes determine an involution on the curve. But the plane joining BC to T is conjugate to the plane joining YZ to X, since T is the pole of the plane XYZ; wherefore the plane BCX contains the pole of the plane YZT. By similar reasoning the pole of the plane YZT lies on each of the six planes joining X to a pair of the points A, B, C, D. This pole is therefore at X. Similarly, Y, Z are the poles of the planes ZXT, XYT, respectively, and X, Y, Z, T are a self-polar tetrad.

A cubic curve which contains four points forming a self-polar tetrad in regard to a quadric Σ will be spoken of as *outpolar* to Σ. It contains not only other self-polar tetrads, as has been proved, but is such that it is met by any quadric, S, which is outpolar to Σ, and does not contain the curve, in six points, say H, forming what has been called a self-conjugate hexad in regard to Σ (Chap. I, p. 47), that is, six points such that the plane containing any three of them is conjugate, in regard to Σ, to the plane containing the other three. A descriptive proof of this may be given. But we may argue as follows: Since there are three independent quadric surfaces containing the curve, all of which contain the points of any self-polar tetrad of points of the curve, we have, through the six points H, four quadrics all of which are outpolar to Σ, namely the three quadrics containing the curve and the quadric S. The equations of these four quadrics are linearly independent, since S does not contain the curve; and, as there are ten terms in the equation of a quadric, the number of quadrics through six points, with equations which are linearly independent, is just four. The number could be more than four only if every quadric described through five, or fewer, of these points, necessarily passed through the others (as a plane described through three of four coplanar points necessarily passes through the other). The six points H have not this character, since, else, a degenerate quadric consisting of a plane through three of the points and a plane through two others would necessarily pass through the sixth. Thus we can infer that every quadric passing through these points H is outpolar to the quadric Σ; in particular a pair of planes containing these six points form together such an outpolar quadric, that is they are conjugate in regard to Σ. Conversely, of such a self-conjugate hexad of points upon the cubic curve, five points may be taken arbitrarily. In particular, if we take five points upon the curve, X, Y, Z, T, U, of which T, U are the extremities of the chord of the curve drawn through the pole of the plane XYZ, in regard to the quadric Σ, then every other point of the curve forms with these five a self-conjugate hexad;

thence it follows that the pole of the plane containing any three of the points X, Y, Z, T, U lies on the line joining the other two. These points constitute then what is called a self-conjugate pentad.

We have defined the cubic as outpolar to Σ when it contains four points forming a self-polar tetrad of Σ. It can, however, be shewn that if the cubic contains the points of a self-conjugate hexad (or of a pentad), it contains also the points of a self-polar tetrad. This is obvious from what was shewn in Chap. I (p. 51); for, thence, the four linearly independent quadrics containing the six points will all be outpolar to Σ; in particular, the three quadrics passing through the curve. Also, it may be remarked, in the notation of p. 47, that if A, B, C, D, E, F be a self-conjugate hexad of points of the curve, the quadric $\Phi(A, B, C, D)$ is that quadric containing the curve which contains the chord EF and is outpolar to Σ.

Ex. 28. A cubic curve, θ, consists of the points common to three quadrics of which every two have a common generator, and is outpolar to a quadric Σ when this last is inpolar to the three former quadrics containing the curve. Similarly a cubic developable, Θ, consists of the planes common to three quadrics regarded as aggregates of their tangent planes, of which every two have a common generator, and the cubic developable may be spoken of as inpolar to a quadric S when this last is outpolar to the three tangential quadrics defining the developable. When Σ consists of the tangent planes of S, we may speak of the curve θ as outpolar to the developable Θ. This case has already arisen in a preceding example (Ex. 16); if an involution of sets of four points be taken on θ, each of these sets forms a self-polar tetrad in regard to a certain quadric, and the planes containing threes of the four points are those of a developable Θ.

Ex. 29. If A, B, C, D and A', B', C', D' be two tetrads of points both self-polar in regard to a quadric, the cubic curve, described to contain A, B, C, D and C', D', has $A'B'$ for a chord, meeting this in two points, P, Q, which are harmonic conjugates in regard to A' and B'. For, by what we have seen, the polar plane of D', in regard to the quadric, which passes through C', meets the curve again in points, P and Q, such that D', C', P, Q are a self-polar tetrad. Thus P, Q lie on the join of A', B' and are harmonic in regard to these.

The equation of the general quadric through A, B, C, D, C', D' is of the form $\lambda U + \mu V + \nu W + \rho\Phi = 0$, where $U = 0, V = 0, W = 0$ may be taken to be independent quadrics containing the cubic curve through these six points. As these three quadrics contain P and Q, we see that the quadrics through the six points meet the line PQ in pairs of points of an involution, of which P, Q are one pair. One such quadric, therefore, contains both A' and B'. Thus, there are three independent quadrics, containing the eight points $A, B, C, D, A', B', C', D'$, any quadric containing A, B, C, D, B', C', D' necessarily passing also through A'.

Any seven general points may be regarded as being seven points from two tetrads which are both self-polar in regard to a properly chosen quadric. For a quadric can be found having four points as a self-polar tetrad, in regard to which every two of three other arbitrary points are conjugate to one another; the pole of the plane of these three, in regard to this quadric, then forms a second self-polar tetrad.

Thus we infer that if two tetrads, of points of general position, be both self-polar in regard to a quadric, any two tetrads formed from these eight points are also two self-polar tetrads in regard to a properly chosen quadric. Also, that all quadrics passing through seven general points pass also through an eighth point. This eighth point may be found by drawing the cubic curve through any six of the seven given points, and then the chord to this curve from the seventh point, and taking thereon the harmonic conjugate of the seventh point in regard to the two points where the chord meets the curve. Or, if the seven given points be A, B, C, D, B', C', D', the eighth point is A', constructed as above, as the pole of the plane $B'C'D'$ in regard to the quadric for which A, B, C, D is self-polar and every two of B', C', D' are conjugate.

In another phraseology, if we have two self-polar tetrads in regard to a quadric, the cubic curve containing any six of the points is outpolar to the point-pair constituted by the remaining two, regarded as a degenerate tangential quadric. The squares of the left sides of the equations of the eight points are in fact connected by a linear relation (cf. Chap. I, p. 51).

Eight points which are such that all quadrics through seven of them necessarily pass through the remaining one, are said to be eight *associated* points. Some particular properties of such a set are given below (p. 154).

Ex. 30. Prove that the squares of the left sides of the equations of eight points of a cubic curve are connected by a linear relation of the form $\Sigma c_r P_r^2 = 0 \; (r = 1 \ldots 8)$. See, for instance, *Proc. Lond. Math. Soc.* IX, 1910, p. 174. For the theory of outpolar cubics, see also Wakeford, *ibid.*, XXI, 1922, p. 101, and *ibid.* p. 114. For the descriptive theory, see Telling, *Proc. Camb. Phil. Soc.*, XXI, 1922, p. 249.

Ex. 31. Consider a focal system (p. 60), and a quadric, such that two particular lines which are polars in the focal system are also polars of one another in regard to the quadric. Let P_1, P_2 be any two points of which the focal planes, respectively ϖ_1, ϖ_2, intersect in the line P_1P_2. Denote by δ the interval P_1P_2 measured with respect to the quadric, in the manner employed for a conic in Vol. II, p. 168, and by θ the angular interval between the planes ϖ_1, ϖ_2, similarly measured with respect to the tangent planes drawn from the line P_1P_2 to the quadric. Consider $\sin\theta/\sin\delta$, expressing this when the focal system and quadric are given by

$$n + kn' = 0,\ Ax^2 + By^2 + Cz^2 + Dt^2 = 0.$$

Shew that the quadric can be so taken that $\sin\theta/\sin\delta$ is independent of P_1 and P_2 (cf. Lie-Scheffers, *Berührungstransformationen*, 1896, p. 233. The corresponding theorem given by Lie for a non-linear complex, *ibid.*, p. 308, is not true).

The rational curve of order n in space of n dimensions. A conic in a plane, and the cubic curve in three dimensions, are evidently both particular cases of a curve in space of n dimensions, of which the coordinates, $x_0, x_1, \ldots, x_n$, of any point of the curve, are respectively proportional to $(n+1)$ integral polynomials in a variable parameter, say θ. We do not enter into the general theory here; but it seems well to remark that any curve in space of r dimensions, in which $r < n$, of which the coordinates of any point are expressible as polynomials of order n in a variable parameter, θ, may be regarded as arising, by projection, from this curve of order n in space of n dimensions; which is thus of fundamental importance. For instance, to illustrate a general proof of this theorem, consider a curve in a plane of which the coordinates, x, y, z, of any point, are expressible as cubic polynomials in θ. Take a point (X, Y, Z, T), in space of three dimensions, of which X, Y, Z are precisely the same functions of θ as are x, y, z, while T is an arbitrary cubic polynomial. This point (X, Y, Z, T) then describes a cubic curve in three dimensions. But to any point of this curve there follows by the equations $x/X = y/Y = z/Z$, that is by projection from the point $(0, 0, 0, 1)$, a corresponding point of the given plane curve. This plane curve has, therefore, a double point, lying on the chord of the space cubic which can be drawn to it from $(0, 0, 0, 1)$; but the space cubic has no double point. For, if it had, a plane, drawn through this double point and two arbitrary points of the curve, would meet the curve in four points, and, therefore, contain it entirely. By a similar argument the curve in space of n dimensions is without double points.

It is supposed in general, in speaking of a curve which is representable rationally by a parameter, that not only is one point of the curve given by each value of the parameter, but, conversely, only one value of the parameter corresponds to a general point of the curve. The parameter can therefore be defined as a rational function of the coordinates of the point of the curve. It can be shewn that if a parameter has been chosen for which this converse expression is not possible, another parameter, a rational function of the former, can be substituted for which this is possible (cf. Vol. II, p. 136. The several values of the unsuitable parameter which correspond all to the same point of the curve are sets of a general involution).

It is clear that the number of points of the general rational curve of the nth order of which the coordinates satisfy a homogeneous equation of the rth order in the $(n+1)$ coordinates, is nr. It is characteristic of the curve that these can be arbitrarily taken. For instance, there are $2r$ points of a conic lying upon a general curve in the plane whose equation is given by a homogeneous equation of order r in the coordinates; but, conversely, such a curve of order r can be drawn through any $2r$ points taken upon the conic, as may be readily verified. Or, again, a surface of equation of order r can be drawn through any $3r$ points taken arbitrarily upon the cubic curve in space of three dimensions, not entirely containing the curve. For instance, a quadric surface can be drawn through any six points of the curve, not entirely containing the curve; or a cubic surface, not entirely containing the curve, through any nine points.

Ex. 1. Find the most general relation, homogeneously of the third order in x, y, z, t, which is identically satisfied by $x=\theta^3$, $y=\theta^2$, $z=\theta$, $t=1$.

Ex. 2. Shew that if a general cubic surface, expressed by the vanishing of a homogeneous relation of the third order in x, y, z, t, be put through two non-intersecting generators of a quadric surface, the remaining intersection is a curve meeting an arbitrary plane in four points, of which the coordinates of a point are expressible rationally by a parameter. Shew, also, how to define, by intersections of manifolds in space of four dimensions, a rational curve of order four which projects into this curve.

Ex. 3. Shew that the equations, for x, y, z, t in terms of a parameter θ,

$$(p\theta+a)x=\theta x_2+ax_1,\ (p\theta+b)y=\theta y_2+by_1, \ldots, (p\theta+d)t=\theta t_2+dt_1$$

represent a quartic curve containing the points of reference and the two points (x_1, y_1, z_1, t_1), (x_2, y_2, z_2, t_2); and that it can be made to pass through two further arbitrary points by suitable choice of the constants a, b, c, d, p. In how many ways is this possible? Find a quadric surface, and a cubic surface containing this curve.

Ex. 4. Shew that a conic, regarded as a curve in three dimensions, depends upon eight parameters. In particular, given seven lines of which no two intersect, which have a common transversal, prove that there exists a single conic which meets every one of the seven lines and also meets the transversal. Analytically, we may regard the conic as the intersection of two quadric cones whose vertices are on the transversal, which touch along this line; we are thus led to seven linear homogeneous equations in eight unknown coefficients. (See Ex. 7, p. 195, below.)

It may be proved that there are ninety-two conics meeting eight lines of general position. See J. Lüroth, *Crelle*, LXVIII (1868), and C. Hierholzer, *Math. Annal.* II (1870).

The curve of intersection of two general quadric surfaces. If two quadric surfaces have a common self-polar tetrad of points, their curve of intersection is essentially different from the cubic curve which has been considered. The coordinates of its points cannot be expressed rationally in terms of a parameter; they can be expressed rationally in terms of *two* parameters, of which one is the square root of a quartic, or cubic, polynomial in the other, or, what is the same thing, if x, y, z, t be the coordinates of a point of the curve, a cubic curve can be found in a plane, in which the coordinates are ξ, η, ζ, this curve being expressed by the vanishing of a homogeneous cubic polynomial in ξ, η, ζ, such that the ratios of x, y, z, t are rational functions of the ratios of ξ, η, ζ, and the ratios of ξ, η, ζ are rational functions of the ratios of x, y, z, t. Such a cubic curve is obtained by projecting the space curve on to a plane from any point of itself. Further, this space curve has the property that, of its intersections with an arbitrary surface, not all can be arbitrarily taken, one of them being always determined when the others are assigned. For instance, the curve meets an arbitrary plane in four points; but, of these, one is determined by the plane containing the other three. Or again, it can be shewn that the curve meets a general quadric surface in eight points (see below); and that the quadric surfaces through seven of these all pass through the remaining one, which, therefore, cannot be taken arbitrarily upon the curve when these seven are given.

Ex. 1. If two quadric surfaces be expressed by the respective equations

$$x^2 + y^2 - z^2 = t^2, \quad ax^2 + by^2 - cz^2 = t^2,$$

any point upon the former is given in terms of θ and ϕ by the expressions $x = \theta + \phi$, $y = 1 - \theta\phi$, $z = 1 + \theta\phi$, $t = \theta - \phi$; if this point lie also on the latter there is a (2, 2) relation connecting θ and ϕ, by means of which either is expressible rationally in terms of the other, and the square root of a quartic polynomial in this other. The systematic consideration of this relation is best undertaken in close connexion with the theory of elliptic functions, and we do not enter upon this here.

Ex. 2. Prove that the quartic curve in space now being considered meets in two points every generator of any quadric surface which contains the curve. Further, that if a variable plane be drawn through two points of the curve, the join of its two other intersections with the curve describes a quadric surface containing the curve. Further, that such a curve can be drawn through eight general points of space. Also, that two chords of the curve can be drawn through an arbitrary point, not lying thereon; and that no line exists meeting the curve in three points.

Ex. 3. Shew that the algebraic determination of the common self-polar tetrad of two general quadrics can be carried out as for the common self-polar triad of two conics given p. 141 of Vol. II; and that the theorem for the reality of the roots of the determining quartic equation, given for the cubic on p. 165 of Vol. II, remains also true.

The determination of a quadric by given conditions. The equation of a quadric contains ten terms; that the quadric should pass through a point of assigned coordinates imposes one linear condition for the coefficients in its equation. That the quadric should pass through several, say through r, given points, requires r such linear conditions for the ten coefficients. If these conditions are independent there are $10-r$ of the coefficients in terms of which the remaining ones can be linearly expressed; then the equation of the general quadric through the r assigned points is of the form $\Sigma\lambda_s U_s = 0$, with $s = 1, \ldots, 10-r$, wherein the $\lambda_1, \lambda_2, \ldots$ are arbitrary, but $U_1 = 0$, $U_2 = 0$, ... represent definite quadrics passing through the assigned points. When this is so the assigned points are called *independent*. It may happen however that, of the r linear conditions for the ten coefficients, one or more is obtainable as a linear aggregate of the others; a very simple example would be that in which four of the assigned points were in line; then a quadric containing any three of these would contain the line, and the condition that it contained the fourth point would be a consequence of the conditions that it contained these three. Algebraically, as the equations are linear, the equation expressing the fourth condition would be a linear function of the equations expressing the three others. These considerations have their application chiefly when the assigned points are already conditioned in some way, as by lying on a curve, or a surface.

For instance, upon the curve of intersection of two quadric surfaces there must be sets of eight points furnishing independent conditions for quadrics required to pass through them; there cannot be nine such independent points; for, by definition, such nine points entirely determine a quadric, whereas, in the present case, if $U=0$, $V=0$ be the given quadrics, the quadric $U+\lambda V=0$ contains all points of their common curve, in particular the nine points, and can yet, by choice of λ, be made to pass through a further arbitrary point. But there must be as many as eight independent points upon the curve; for if, for example, there were but seven, this would mean that every quadric through these seven necessarily contained all points of the common curve; it would then be possible to describe a quadric through this curve and two other arbitrary points. We can, however, prove that only one quadric can be passed through this curve and a further single point, O; namely by considering the section of the quadrics with a general plane drawn through O. In this argument we have spoken of a quadric passing through all the points common to two given quadrics, as if this were a description quite free of ambiguity; when the given quadrics have multiple points of intersection, as for example in the case when they touch at all points of a plane conic, the description needs interpretation. This is furnished by first agreeing as to what is meant by a conic passing through all the points common to two given conics, even when there are coincidences among these; and then speaking of a quadric as passing through all points common to two given quadrics when it meets every plane in a conic which passes through all the points common to the sections, by this plane, of the two given quadrics. It will be seen that this meets all cases.

Ex. 1. The necessary and sufficient condition that two quadrics $U=0$, $V=0$ should intersect in two conics, lying in planes $P=0$, $Q=0$, is the existence of an identity of the form $\lambda U+\mu V+\nu PQ=0$, where λ, μ, ν are constants. And this remains true when the two conics coincide, the quadrics touching one another at all points of this.

Ex. 2. Three quadric surfaces, which have no common curve, meet in eight points. Taking for points of reference a self-polar tetrad for one of the quadrics, the common points may be found by substituting, in both the other two quadric equations, the coordinates $x=\theta+\phi$, $y=1-\theta\phi$, $z=1+\theta\phi$, $t=\theta-\phi$. The result is two equations, $(2, 2)=0$ and $[2, 2]=0$, which are both polynomials of the second order in each of θ and ϕ. The elimination of ϕ between these leads to an equation for θ by the solution of which the intersections can be found. Geometrically, regarding $(\theta, \phi, 1)$ as coordinates in a plane, the statement is that two quartic curves in a plane, which have common two double points, have eight other common points. The result may be regarded as always true, by proper interpretation of multiple intersections.

Ex. 3. Every quadric through seven points of space of which no four lie in a plane, and, therefore, no three lie in a line, has an equation of the form $\lambda U+\mu V+\nu W=0$, wherein U, V, W can be supposed to be of the respective forms

$$yz+(a_1x+b_1y+c_1z)\,t,\ zx+(a_2x+b_2y+c_2z)\,t,\ xy+(a_3x+b_3y+c_3z)\,t.$$

We can now prove that $U=0$, $V=0$, $W=0$ have eight common points, so once more proving that all quadrics through seven given points have an eighth point common. In fact $V=0$, $W=0$ both contain the line $x=0$, $t=0$; thus they also intersect in a cubic curve, meeting this line in two points. This curve will meet $U=0$ in six points; these, with the two points in which $U=0$ is met by the line $x=0$, $t=0$, are the eight points in question, and, of these six, two are on the line $y=0$, $z=0$. If we put $x=\theta t$ in $V=0$, $W=0$, we find $a_2x+b_2y+(c_2+\theta)z=0$, $a_3x+(b_3+\theta)y+c_3z=0$, from which we have the expression of the cubic curve by the parameter θ. Substituting this expression in $U=0$, the determination of the four points of intersection of the three quadrics, beside the given points of reference (1, 0, 0, 0), (0, 1, 0, 0), etc., is reduced to the solution of a quartic equation.

Upon the curve of intersection of two general quadrics, there can be taken such seven independent points as we have supposed. Thus, the curve of intersection of two general quadrics is such that, of its intersections with another quadric, one is determined by the assignment of the other seven.

Ex. 4. Shew that if three quadric surfaces have a line in common, they have four common points not lying on this. And, if three quadric surfaces have a conic in common, they have two common points not lying thereon.

Ex. 5. If all possible quadric cones be constructed which contain six arbitrary fixed points of general position, prove that the vertices of these cones lie upon a surface meeting an arbitrary line in four points (Weddle's surface). Shew that this surface contains the fifteen lines joining the pairs of the six given points, and the ten lines of intersection of the pairs of planes each containing three of the points; and contains also the cubic curve passing through the six given points. Any line drawn through one of the six given points meets the surface in only two other points.

Ex. 6. The vertices of the quadric cones containing seven arbitrary fixed points of general position lie upon a curve which meets an arbitrary plane in six points. This curve has for chords the twenty-one joins of the seven points, and also the seven joins of these points to the further point common to all the quadrics containing these seven given points. (Hesse, *Crelle's J.*, XLIX, 1855, p. 279.)

Theorems for eight associated points. Some results may be given in regard to a set of eight points which are such that every quadric through seven of them also passes through the eighth.

Ex. 1. If A, B, C, M, D', L be six points of a conic it follows, from Pascal's theorem, that the point (CM, LA) is on the line joining the point (AB, MD') to the point $(BC, D'L)$.

Now let A, B, C, D be any points of a quadric surface, of which l and m are two generators of the same system. Let these generators meet the section of the quadric by the plane ABC respectively in

L and M, and let the generator of the quadric, of the opposite system, which passes through D, meet this section in D'. The point (CM, LA) is then on the line of intersection of the planes Cm, lA; the point (AB, MD') is the point where the line AB meets the plane mD, and may be described as the point (AB, mD); so the point $(BC, D'L)$ is the point (BC, Dl). Thus the line (Cm, lA) meets the join of the points (AB, mD), (BC, Dl).

Ex. 2. Let A, B, C, D and A', B', C', D' be any general set of associated points. There are then three linearly independent quadrics passing through these, and a quadric can be put through A, B, C, D to have $A'B'$ and $C'D'$ as generators of the same system. Denoting the points, respectively, by the numbers 3, 4, 5, 8 and 1, 2, 6, 7, and identifying the lines l, m of Ex. 1 respectively with the generators 12, 67, we thus reach the result that, if 1, 2, 3, 4, 5, 6, 7, 8 be a general set of associated points, the line (123, 567) intersects the join of the two points (34, 678), (45, 812).

Ex. 3. If we then remark that this join lies on the plane 345, we see that this join is also met by the line (345, 781). Again, the line (234, 678) contains the point (34, 678), and the line (456, 812) contains the point (45, 812). Thus the four lines

(123, 567), (234, 678), (345, 781), (456, 812)

are all met by the line joining the points, (34, 678), (45, 812). These four lines are, however, only interchanged among themselves by cyclical interchange of the numbers; it follows that they are met also by the join of the points (45, 781), (56, 123), and so on, in all by eight such joining lines. Thus the four lines are generators of the same system of a quadric.

Ex. 4. Let 2, 3, 4, 5, 7, 8 be any six points, and 1 a further point. From the last point let the transversal be drawn to the lines 23, 57, meeting these, respectively, in P and P'. Similarly, let Q, on 78, and Q', on 34, be on a line through 1; and R, on 45, with R', on 82, be also on a line through 1. The lines QR', RP', PQ' are then generators of a quadric, of which $Q'R$, $R'P$, $P'Q$ are also generators. This quadric meets the lines 23, 34, 45, 57, 78, 82, respectively, in other points P_1, Q_1', R_1, P_1', Q_1, R_1'. It can be shewn that the three lines P_1P_1', Q_1Q_1', R_1R_1' meet in a point. Denoting this by 6, the points 1, 2, ..., 8 are eight associated points. The proof follows easily from Ex. 2.

Ex. 5. For the determination of the eighth point of an associated set when seven are given, we have, by what precedes, two theorems: (1), the cubic curves determined by the two sets of six points, A, B, C, D, A', C'; A, B, C, D, B', C' have, respectively, chords meeting in the eighth point, D', required. The chord of the first curve is that drawn from B', that of the second curve is that drawn from A'. Or, as we have seen (Ex. 18, p. 139, above) that a single cubic curve exists passing through five given points, having a given line as chord, we can regard D' as the common point of the various cubic curves constructed each to pass through five of the seven given points and have the join of the other two as chord. (2), if A, B, C, D be a self-polar tetrad in regard to a quadric, and every two of A', B', C' be conjugate in regard to this quadric, the point D' is the pole of the plane $A'B'C'$ in regard to this. Hesse has given a linear construction for D' founded on this last fact (see *Ges. Werke*, pp. 46, 48, 680. See also Cayley, *Papers*, I, p. 425 (1849), for another view of the problem); it depends upon a theorem of plane geometry: In a plane let A, B, C be a self-polar triad in regard to a conic, and P, U; Q, V; R, W be three pairs of conjugate points in regard to this conic; then the polar lines, respectively l, m, n, of U, V, W, can be linearly constructed. Let BU, AV meet in O; let PO meet AC in Y, and QO meet BC in X; further, let XV meet AQ in H, and YU meet BP in K. Also let AC meet HK in L, and BC meet HK in M. Then PL is the polar of U, and QM the polar of V. Similarly, the line n can be constructed.

This being assumed, and the points A, B, C, D, A', B', C' given as above, let the six lines $DA', B'C', DB', C'A', DC', A'B'$ meet the plane ABC, respectively, in U, P, V, Q, W, R. The triad ABC will be self-polar in regard to the conic in which the quadric meets the plane ABC, and the polar line of DA' will be a line, l, lying in this plane and in a plane which contains $B'C'$. Thus the polar line of U in regard to this conic will be the line l, and this will contain P. Similarly V, Q will be conjugate points in regard to this conic, the polar line, m, of V containing Q; and W, R will be such that the polar line, n, of W, contains R.

The lines l, m, n being constructed as explained, the point D', required, is the intersection of the planes $(l, B'C')$, $(m, C'A')$, $(n, A'B')$.

Ex. 6. If the curve of intersection of two general quadric surfaces be projected on to a plane from any point of itself, it becomes a cubic curve. Conversely, it can easily be shewn that a curve in a plane, expressed by the vanishing of a homogeneous cubic polynomial in the three coordinates, may be regarded as arising by the projection of the curve of intersection of two quadric surfaces. Bearing in mind that the latter curve meets every generator, of a quadric surface on which it lies, in two points, the theorem that seven points in space determine an eighth is at once seen to be in correspondence with the theorem, for a plane, that all general cubic curves with eight common points pass through a ninth. We may construct this ninth point as follows: let A, B, C, M, N and P, Q, R be eight points of a plane, and D a point not lying in this plane; let Ω be a quadric containing A, B, C and the lines DM, DN (other than the plane pair ABC, DMN); let DP, DQ, DR meet this quadric (beside at D) in P_1, Q_1, R_1. Now let the cubic curve in space, which contains the five points D, A, P_1, Q_1, R_1 and has BC for a chord, meet Ω again in O_1. Then DO_1 meets the plane in the ninth point required. For this problem see also Cayley, *Papers*, IV, pp. 495–504 (1862). Also p. 178, below.

Ex. 7. A further way of regarding the theorem that quadrics through seven points have an eighth common point should be referred to briefly. Taking five arbitrary points, of which no four lie in a plane, there are five linearly independent quadrics passing through these, say $U_1 = 0, U_2 = 0, \ldots, U_5 = 0$. If four of the points be the points of reference and the fifth be $(1, 1, 1, 1)$, the functions $U_1, \ldots, U_5$ are in fact arbitrary linear functions of the six functions

$$\xi = y(t-z), \quad \eta = z(t-x), \quad \zeta = x(t-y),$$
$$\xi' = (t-y)z, \quad \eta' = (t-z)x, \quad \zeta' = (t-x)y,$$

which are equivalent to five in virtue of the identity

$$\xi + \eta + \zeta = \xi' + \eta' + \zeta',$$

and are further connected by the relation $\xi\eta\zeta = \xi'\eta'\zeta'$. We may regard the ratios of any five independent linear functions of $\xi, \eta, \zeta, \xi', \eta', \zeta'$ as coordinates of a point of space of four dimensions. Thereby, to any point of our original three-fold space, other than certain exceptional points for which all of $\xi, \eta, \ldots, \zeta'$ are zero, there corresponds a point of the three-fold locus, Σ, lying in the four-fold space, for which $\xi\eta\zeta = \xi'\eta'\zeta'$; conversely, save for exceptional points, to any point of this three-fold locus, Σ, there corresponds the point of the original space given by

$$x/t = (\eta' - \zeta)/(\xi - \xi'), \quad y/t = (\zeta' - \xi)/(\eta - \eta'), \quad z/t = (\xi' - \eta)/(\zeta - \zeta').$$

To the general quadric of the original space which passes through the given five points there corresponds, then, in the four-fold space, the intersection of the locus Σ with the locus expressed by an equation of the form $A\xi + B\eta + C\zeta = A'\xi' + B'\eta' + C'\zeta'$, wherein the relation $\xi + \eta + \zeta = \xi' + \eta' + \zeta'$ is understood. Let such a three-fold locus, or space, expressed by a linear equation, be, for distinctness, denoted by σ. To the quadrics of our original space passing through seven given points, consisting of the five chosen fundamental points and two other points, say A and P, will correspond the spaces σ, in the four-fold space, which pass through two given points, say $(\alpha, \beta, \gamma, \alpha', \beta', \gamma')$ and $(\lambda, \mu, \nu, \lambda', \mu', \nu')$, of the locus Σ. These will contain all points of the line joining these two points of Σ; and, as we see by substituting expressions of the forms $\alpha + \phi\lambda, \ldots, \gamma' + \phi\nu'$ for the coordinates in the equation of Σ, there is a third point of Σ lying on this line, through which, then, all these spaces σ pass. This is the point corresponding to the eighth intersection of the quadrics in the original space which pass through the seven given points.

The correspondence can be followed in detail. We limit ourselves to the result in the four-fold space which corresponds to the fact that the eighth point lies on the chord, drawn from P, to the cubic curve containing the five fundamental points and the point A. Let this point be (a, b, c, d); this cubic curve is then given by

$$x = a(\theta - a)^{-1}, \quad y = b(\theta - b)^{-1}, \quad z = c(\theta - c)^{-1}, \quad t = d(\theta - d)^{-1},$$

for varying values of θ, the points $(1, 1, 1, 1)$, (a, b, c, d) arising respectively for $\theta = 0$, $\theta = \infty$. It is easily seen that the locus in the four-fold space to which this corresponds is the line given by the three simultaneous equations $\xi/\alpha = \xi'/\alpha'$, $\eta/\beta = \eta'/\beta'$, $\zeta/\gamma = \zeta'/\gamma'$. This line passes through the point $(\alpha, \beta, \gamma, \alpha', \beta', \gamma')$, and lies entirely on the locus Σ. Through this line, and the point $(\lambda, \mu, \nu, \lambda', \mu', \nu')$, corresponding to P, there passes a plane; this plane meets the locus Σ, in fact, in a conic, as we see from the fact that any line meets the locus Σ in three points; the equations of the conic may easily be found. This conic is met by the line, given by the equations $\xi/\alpha = \xi'/\alpha'$, etc., in two points; it corresponds to the chord drawn from P in the original space, to meet the cubic curve; by its construction it contains the point $(\lambda, \mu, \nu, \lambda', \mu', \nu')$, and the line joining $(\alpha, \beta, \gamma, \alpha', \beta', \gamma')$ to this point meets the conic again in the point of Σ which corresponds to the eighth intersection of the quadrics in the original space.

The locus Σ will be found to be of great interest to us. If, instead of $\xi, \eta, \zeta, \xi', \eta', \zeta'$, we use coordinates

$$x_1 = \tfrac{1}{2}(\eta + \zeta - \xi), \qquad x_2 = \tfrac{1}{2}(\zeta + \xi - \eta), \ \ldots,$$
$$x_4 = \tfrac{1}{2}(\xi' - \eta' - \zeta'), \ \ldots, \ x_6 = \tfrac{1}{2}(\zeta' - \xi' - \eta'),$$

its equations take the forms $\Sigma x_r^3 = 0$, $\Sigma x_r = 0$, $r = 1, \ldots, 6$.

The set of eight associated points was considered by von Staudt, *Beiträge zur Geometrie der Lage*, III, 1860, pp. 372 ff. The theorem of Ex. 2, above, was given by Hesse; see *Ges. Werke*, p. 680, and various papers in *Crelle's J.* The theorem of Ex. 3 was given by Zeuthen; see *Acta Math.* XII, 1889, p. 363. See also Dobriner, *ibid.* p. 339; and Schroeter, *Acta Math.* XIV, 1890. The locus Σ considered in Ex. 7 was studied by Segre; see *Atti Acc. sc. Torino*, XXII, 1886, 7; p. 547.

CHAPTER IV

THE GENERAL CUBIC SURFACE; INTRODUCTORY THEOREMS

In this book we regard the theory of a cubic surface as best treated in connexion with a certain figure in space of four dimensions. But the preliminary theorems for a general cubic surface are so intimately related with other results included in this volume that it seems desirable to give some account of them.

The theorem of a double-six of lines. It is convenient to give at once an independent proof of a theorem which arises implicitly below. Suppose that a, b, c, d, e are five lines, no two meeting one another, which have a common transversal, f'. There is then, beside f', a common transversal of every four of the five lines. The theorem referred to is that the five transversals so arising have themselves a common transversal. Denote by e' the transversal of a, b, c, d, beside f', by d' the transversal of a, b, c, e beside f', and so on. We are to prove that the five lines a', b', c', d', e' are all met by a line. It is clear that no two of these lines intersect; if, for instance, a' and b' were in one plane, every two of c, d, e, all of which meet a' and b', would intersect.

We prove that a', b', c', d', e' have a common transversal by shewing that the common transversal of a', b', c', d', other than e, coincides with the common transversal, other than d, of a', b', c', e'. This we do by shewing that the common transversal, other than e, of a', b', c', d', meets a', b', c', respectively, in three points which can be constructed from the lines a, b, c, a', b', c', f' only. In fact, denoting the nine intersections,

$$(b', c),\ (b, c'),\ (c', a),\ (c, a'),\ (a', b),\ (a, b'),\ (a, f'),\ (b, f'),\ (c, f'),$$

respectively by

$$A,\ A',\ B,\ B',\ C,\ C',\ U,\ V,\ W,$$

we shew that the common transversal of a', b', c', d', other than e, meets a' on the plane (A, A', U), meets b' on the plane (B, B', V), and meets c' on the plane (C, C', W).

Let f denote the common transversal, other than e, of a', b', c', d', and let this meet the lines a', b', c', respectively, in X, Y, Z. Consider the two quadric surfaces (b, c, e) and (b', c', d'), having, respectively, as generators of the same system, the two triplets of lines indicated. As the line e meets b', c', d', it is a generator of the second quadric as well as of the first; as d' meets b, c, e, it is a generator of the first quadric as well as of the second. As these two quadrics have thus two intersecting lines as common generators, their remaining common points lie in a plane. Of such points, not lying on e or d', are the points A, or (b', c), and A', or (b, c'); so also the point U, or (a, f'), is such a point, since the line a, meeting b', c', d', is a generator of the second quadric, and f', meeting b, c, e, is a generator of the first quadric. Lastly, the point X, or (a', f), is common to the two quadrics, since a' meets b, c, e, and f meets b', c', d'. Thus X, where f meets a', lies on the plane (A, A', U). By considering the two quadrics (c, a, e) and (c', a', d'), we similarly shew that Y lies on the plane (B, B', V); and, by considering the two quadrics (a, b, e) and (a', b', d'), that Z lies on (C, C', W).

For the common transversal, other than d, of a', b', c', e', we proceed in the same way, using e', d instead of d', e, respectively. The planes (A, A', U), (B, B', V), (C, C', W) arise once more. The theorem is thus proved.

Such an aggregate of twelve lines

$$a,\ b,\ c,\ d,\ e,\ f,$$
$$a',\ b',\ c',\ d',\ e',\ f',$$

consisting of two sextuples, wherein each line meets five lines of the sextuple to which it does not belong, but is skew to the other lines of its own sextuple, and is also skew to a corresponding line of the other sextuple, is called a double-six. We shall commonly speak of the lines of the double-six which form a sextuple as a *row* of the double-six.

From the twelve lines of the double-six, fifteen other lines can be defined, as follows: associating the lines $a, b, \ldots, f, a', b', \ldots, f'$, respectively, with the numbers $1, 2, \ldots, 6, 1', 2', \ldots, 6'$, let the line of intersection of the plane of a, b', with the plane of a', b, say, of the plane $[1, 2']$ with the plane $[1', 2]$, be denoted by c_{12}; there will be fifteen such lines, obtained by taking every pair of the numbers $1, 2, \ldots, 6$. Fifteen other lines can also be obtained from the double-six: we may join the point of intersection, $(1, 2')$, of the lines a, b', to the point of intersection, $(1', 2)$, of the lines a', b; this construction can be made in fifteen ways. We may denote the joining lines by symbols c_{rs}'.

It has incidentally been shewn above that the four points $(2, 3')$, $(2', 3)$, $(1, 6')$, $(1', 6)$ lie in one plane, this being the fact that the points A, A', U, X are in one plane. Thus the line c_{23}' intersects the line c_{16}'. By a precisely similar argument it follows that, if no two of the numbers r, s, i, j are the same, the lines c_{rs}', c_{ij}' intersect one another. Thence, by the dual argument, the plane of two intersecting lines being considered instead of their point of intersection, it follows, if no two of the numbers r, s, i, j are the same, that any two lines c_{rs}, c_{ij} intersect one another.

We have now two sets, each of ($12+15$, or) twenty-seven lines, one set consisting of $a, b, \ldots, f, a', b', \ldots, f'$, together with the lines c_{rs}, the other set consisting of the same twelve lines together with the lines c_{rs}'. Let us consider the former set. It is easy to see that the lines of this set are symmetrical as regards their intersections with one another: each line of the twenty-seven intersects ten others of the set, and these consist of five pairs of mutually intersecting lines, so that in the plane of any two intersecting lines of the set is a third line of the set. For instance, the line 1 (or a) is met by the ten lines

$$2', c_{12};\ 3', c_{13};\ 4', c_{14};\ 5', c_{15};\ 6', c_{16},$$

which consist of five pairs of intersecting lines. Similarly, the line c_{12} is met by the lines of the five pairs, each of two intersecting lines,

$$1, 2';\ 1', 2;\ c_{34}, c_{56};\ c_{35}, c_{64};\ c_{36}, c_{45}.$$

Further, it is possible to arrange the twenty-seven lines in two sets of, respectively, twelve and fifteen lines, in other than the original way, so that the twelve lines form a double-six, and the fifteen are deducible from these by the same rule as were the lines c_{rs} from the original double-six; there are in fact thirty-six such arrangements. Writing, for greater clearness, 23 instead of c_{23}, and so on, the following twelve lines

$$\begin{matrix} 1, & 1', & 23, & 24, & 25, & 26 \\ 2, & 2', & 13, & 14, & 15, & 16 \end{matrix}$$

are at once seen to form a double-six; further, the lines formed from these, as were the lines c_{rs} from the original double-six, are easily seen to be the other fifteen lines of the original twenty-seven. Such a double-six as this can be constructed in fifteen ways. Another double-six, typical of twenty double-sixes, is given by

$$\begin{matrix} 1\,, & 2\,, & 3\,, & 56, & 64, & 45 \\ 23, & 31, & 12, & 4', & 5', & 6', \end{matrix}$$

and the remaining fifteen lines arise from this by the same rule as before. We see then how to obtain $1+15+20$, or thirty-six, double-sixes.

Reciprocity of the figure. Considering the original double-six, it can be shewn that there exists a quadric surface in regard to which the six lines $1, 2, \ldots, 6$, of one row, are, respectively, the polar lines of the corresponding lines, $1', 2', \ldots, 6'$, of the other row. In regard to this quadric surface, then, the line c_{rs} is the polar of c_{rs}'. To prove this, consider the fifteen triplets of lines each consisting of two lines of the first row of the double-six taken with that one of the lines c_{rs} which intersects both; such a triplet is given, for instance, by the three lines 1, 2, c_{12}. With these fifteen triplets consider also the similar triplets each formed from two lines of the second row of the double-six taken with that one of the lines c_{rs} which intersects both, such, for instance, as 3′, 4′, c_{34}. Then it can be seen that a definite quadric exists, possibly degenerating into two planes, containing the combination formed by any triplet of the first set of fifteen triplets, taken with any triplet of the second set. For instance, the two triplets 1, 2, c_{12} and 1′, 2′, c_{12} both lie in the plane-pair [1, 2′], [1′, 2], whose line of intersection is the line c_{12}; again, the two triplets 1, 2, c_{12} and 2′, 3′, c_{23} lie on the plane-pair of which one plane contains the lines 1, 2′, c_{12} and the other plane contains the lines 2, 3′, c_{23}; while two triplets such as 1, 2, c_{12} and 3′, 4′, c_{34} consist of three lines 1, 2, c_{34}, of which no two intersect, together with three other such lines, 3′, 4′, c_{12}, of which, however, each intersects all the first three, so that the six lines all lie on a quadric surface. Assuming the statement verified, there will be, in all, 15 . 15 or 225 quadrics; it is then, further, the case that the equations of these quadrics, expressed in point coordinates, are linear functions of *nine* of them. For there are three linearly independent quadrics passing through a triplet consisting of two skew lines and a transversal of these: Let P, Q, R, S denote four triplets each containing two lines of the first row of the double-six, and P', Q', R', S' denote four triplets each containing two lines of the second row of the double-six, and let (S, S') denote the quadric containing the triplets S and S', and so in general; then, if the quadrics (S, P'), (S, Q'), (S, R') be linearly independent, the quadric (S, S') is a linear function of these; if the quadrics (P', P), (P', Q), (P', R) be linearly independent, the quadric (S, P') is a linear function of these; similarly, with a like hypothesis, the quadric (S, Q') is expressible by (Q', P), (Q', Q), (Q', R), and the quadric (S, R') by (R', P), (R', Q), (R', R); finally, every one of the quadrics is thus seen to be expressible by the nine obtained by taking P, Q, R with P', Q', R', provided that, in every case, the three quadrics (S, P'), (S, Q'), (S, R') are linearly independent. A detailed scrutiny shews that this is so in general.

There is, however, a single quadric envelope which is inpolar to nine linearly independent quadrics (above, p. 52). Thus there exists a quadric envelope, in general unique, which is inpolar to the 225 quadrics spoken of. This we may denote by Σ; it will sometimes be spoken of as the Schur quadric associated with the double-six.

Now take the three pairs of triplets

1, 2, c_{12} and 1′, 2′, c_{12}; 1, 2, c_{12} and 2′, 3′, c_{23}; 1, 3, c_{13} and 1′, 2′, c_{12},

lying, respectively, in the three plane-pairs

[1, 2′, c_{12}], [1′, 2, c_{12}]; [1, 2′, c_{12}], [2, 3′, c_{23}]; [1, 2′, c_{12}], [1′, 3, c_{13}];

each of the pairs of planes being conjugate to one another in regard to the quadric Σ, the pole of the plane $[1, 2']$ is seen to be the common point of the planes $[1', 2]$, $[2, 3']$, $[1', 3]$, namely, the point $(1', 2)$. Similarly, the pole of the plane $[1, 3']$ is the point $(1', 3)$. The two planes both contain the line 1, the two points both lie on the line $1'$. Thus the lines $1, 1'$ are polars of one another in regard to the quadric Σ, as are, similarly, 2 and $2'$, 3 and $3'$, etc.

The quadric Σ is in fact determined by the six lines $1, 2, 3, 1', 2', 3'$, only. These six lines are the edges of a skew hexagon, A, B', C, A', B, C', wherein A, A' are, respectively, the points (b', c), (b, c'), or $(2', 3)$, $(2, 3')$, etc., and the quadric Σ is such that opposite edges of this hexagon are polar lines of one another. Thus the point A, the intersection of the edges AB' and AC', is the pole of the plane $A'BC$, etc. Let the three planes $A'BC$, $B'CA$, $C'AB$ meet in the point D, so that AD is the intersection of the planes $AB'C$, $AC'B$, etc. Then D is the pole of the plane ABC, and A, B, C, D are a self-polar tetrad of points in regard to Σ; while A', B', C' are three points of which every two are conjugate to one another in regard to Σ. These two conditions determine the quadric. It will, however, be seen to be of advantage to have considered the quadric from the point of view we have taken.

Another determination of the double-six. It can be shewn that, if we have six lines, a, b', c, a', b, c', forming a skew hexagon, as here, and another line, d, which meets one set of alternate edges, a', b', c', of the hexagon, then a double-six is determined, which will, therefore, be that here discussed with the same notation for the lines. The proof of this is instructive, and may be given. Consider the aggregate of lines which meet the three skew lines a', b', c', these being the generators, of the other system from these, of the quadric determined by a', b', c'. Let us regard two of these lines, say (x) and (y), as *corresponding*, when they are both met by some transversal of the three other alternate edges, a, b, c, of the hexagon. We can then see that to any (x) there correspond two lines (y), and, to any (y), two lines (x); for when a line (x), meeting a', b', c', is taken, there are two transversals of the four lines $(x), a, b, c$; if one of these, say (z), be taken, there is a definite transversal (y), beside (x), of the four lines $(z), a', b', c'$; similarly when (y) is taken first. A line (x), meeting a', b', c', may be defined by a parameter, x, as, for instance, by lying in a definite plane, of equation $U + xV = 0$, passing through a'; the correspondence between the lines (x) and (y) may then be represented by an equation connecting two parameters, x and y; this equation, by the geometrical definition, will be rational, of the second order in each of the parameters, and symmetrical in regard to these. If y_1 and y_2 be the two values of y given by this equation, corresponding to a value of x, it may, or may not, happen, that the two values corresponding to the value, y_1, of x, are x and y_2; this will happen, however, for all values of x if it happen for a particular value. In such case the three values x, y_1, y_2 will be one set of a general involution of triads of values (Vol. II, pp. 135 ff.). If then we can shew that there exists one set of three transversals of a', b', c' with the property that every two of them are met by a transversal of a, b, c, it will follow that, if we take the line d, arbitrarily, to meet a', b', c', and then the lines e, f, also to meet a', b', c', by the condition that e meets the transversal, f', of a, b, c, d, and f meets the other transversal, e', of a, b, c, d, then e and f will also meet a transversal, d', of a, b, c. We shall then have constructed a double-six, clearly unique, from the lines a, b, c, a', b', c', d. It is, however, easy to find one set of three transversals of a', b', c' with the property that every two of them are met by a transversal of a, b, c. For, let O and ϖ be, respectively, a point and a plane which are pole and polar of one another in regard to both the quadrics (a, b, c), (a', b', c'), defined, respectively, by the generators indicated; then, through the point where the line a' meets the plane ϖ, there passes a generator, say a_1', of the quadric (a', b', c'), which meets b' and c', and is the harmonic image of a' in regard to O and ϖ (see above, p. 80); similarly, through the point where the line a meets ϖ, there passes a generator, say a_1, of the quadric (a, b, c), which meets b and c as well as a. With an analogous notation in general, since b' and c' both meet a, it follows that b_1' and c_1', which both meet a', b', c', also both meet a_1. Three transversals of a', b', c' which are such that every two of them are met by a transversal of a, b, c, are, therefore, given by a_1', b_1', c_1'. Thus the proof of the existence of the double-six is completed.

The Steiner trihedral pairs. It has been shewn above that the quadrics containing the pairs of triplets of lines such as r, s, c_{rs} and r', s', $c_{r's'}$ are linearly dependent from nine such quadrics. These may in fact be taken to be those arising from the lines 1, 2, 3, 1′, 2′, 3′, c_{23}, c_{31}, c_{12}, for which the nine quadrics are all plane-pairs. Let the triplets in question, namely

$$2, 3, c_{23};\ 3, 1, c_{31};\ 1, 2, c_{12};\ 2', 3', c_{23};\ 3', 1', c_{31};\ 1', 2', c_{12},$$

be denoted, respectively, by p, q, r, p', q', r', and the planes [2, 3′], [2′, 3], [3, 1′], [3′, 1], [1, 2′], [1′, 2] be denoted, respectively, by α, α', β, β', γ, γ'; the quadrics containing the pairs of triplets

$$q, r';\ q', r;\ r, p';\ r', p;\ p, q';\ p', q;\ p, p';\ q, q';\ r, r'$$

are then, respectively, the plane-pairs

$$\beta, \gamma;\ \beta', \gamma';\ \gamma, \alpha;\ \gamma', \alpha';\ \alpha, \beta;\ \alpha', \beta';\ \alpha, \alpha';\ \beta, \beta';\ \gamma, \gamma'.$$

The nine lines are the intersections of the three planes α, β, γ

with the three planes α', β', γ'. If, as above, we introduce the point D, the intersection of the planes α, β, γ, and the lines DA, DB, DC, joining this to the points $(2', 3)$, $(3', 1)$, $(1', 2)$, respectively, and also introduce the point D', the intersection of the planes α', β', γ', and the lines $D'A'$, $D'B'$, $D'C'$, joining this, respectively, to the points $(2, 3')$, $(3, 1')$, $(1, 2')$, we shall in the aggregate obtain the figure consisting of all the intersections of the six planes α, β, γ,

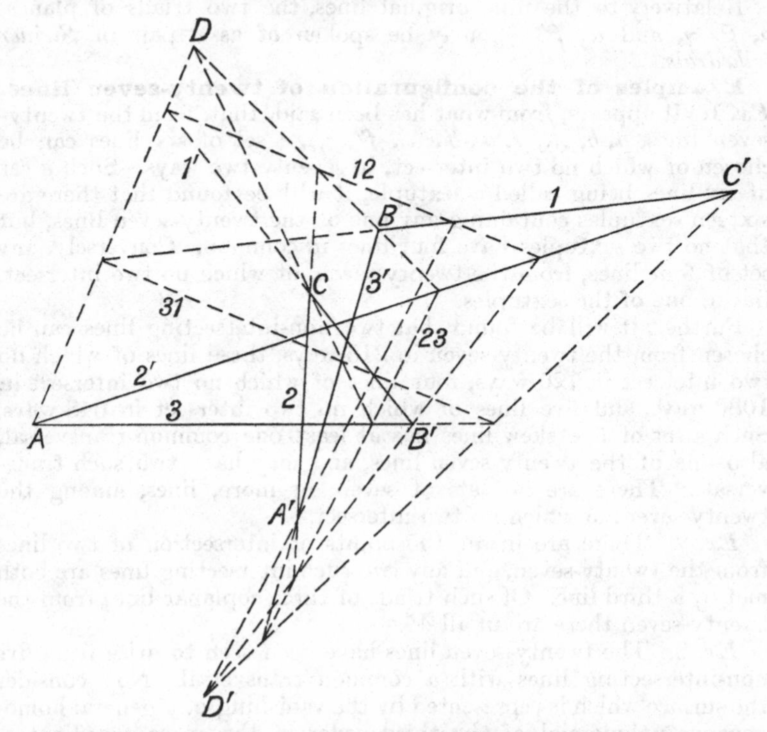

α', β', γ'. This contains twenty points, fifteen lines and six planes, of which each point lies in three of the lines and in three of the planes, each line contains four of the points and lies in two of the planes, while each plane contains ten of the points and five of the lines, so that the incidences can be denoted by the symbol

$$20\,(.\,, 3, 3)\; 15\,(4, .\,, 2)\; 6\,(10, 5, .).$$

The twenty points are the six points A, B', C, A', B, C', the points D, D', and the three sets of four points such as, for example, the intersections of c_{12} with 1, $2'$, $1'$, 2. The fifteen lines are the lines 2, $3'$, ..., c_{12}, originally considered, together with the six such as DA; each of these six lines contains three points of intersection of two of the nine original lines; for instance, DA contains the three points $(2', 3)$, $(1, c_{23})$, $(1', c_{12})$; and the nine lines consist of three coplanar triads all in perspective from D, and consist also of three coplanar triads in perspective from D'. The figure is a section, by a three-fold, of the figure in space of four dimensions considered on pp. 214, 215 of Vol. II.

Relatively to the nine original lines, the two triads of planes, α, β, γ, and α', β', γ', may be spoken of as a pair of *Steiner trihedrals.*

Examples of the configuration of twenty-seven lines. *Ex.* 1. It appears, from what has been said, that, from the twenty-seven lines, $a, b, \ldots, f, a', b', \ldots, f', c_{rs}$, a set of six lines can be chosen of which no two intersect, in seventy-two ways. Such a set of six lines being called a sextuple, it will be found that there are sixteen sextuples containing any one of the twenty-seven lines, but that no two sextuples have four lines in common. Conversely, any set of four lines, from the twenty-seven, of which no two intersect, lies in one of the sextuples.

Further, it will be found that two non-intersecting lines can be chosen from the twenty-seven in 216 ways, three lines of which no two intersect in 720 ways, four lines of which no two intersect in 1080 ways, and five lines of which no two intersect in 648 ways. Such a set of five skew lines has at least one common transversal, also one of the twenty-seven lines, and may have two such transversals. There are no sets of seven, or more, lines, among the twenty-seven, of which no two intersect.

Ex. 2. There are in all 135 points of intersection of two lines from the twenty-seven, and any two such intersecting lines are both met by a third line. Of such triads of three coplanar lines from the twenty-seven there are in all 45.

Ex. 3. The twenty-seven lines have been seen to arise from five non-intersecting lines with a common transversal. Now consider the surface which is represented by the vanishing of a general homogeneous polynomial of the third order in the space coordinates; this is called a cubic surface, being met by an arbitrary line in three points, the substitution of values of the forms $x_0 + \lambda x_1$, $y_0 + \lambda y_1$, etc., for the coordinates, leading to a cubic equation in λ. The conditions that a given line should lie entirely on the surface are four in number, all linear in the coefficients of the equation of the cubic surface; thus the conditions for the cubic surface, that it should contain one given line and five other skew lines which meet this are, in all, nineteen in number. The cubic surface, whose equation contains twenty terms, can thus be determined to contain these six lines. The cubic surface will then equally contain the other twenty-one lines constructed as above from these six, since each additional line is such as to meet four of those already taken.

Conversely, we may expect that any cubic surface, of general type, contains such a set of twenty-seven lines. For the figure of five non-intersecting lines with a common transversal is itself of amplitude nineteen, depends, that is, on nineteen constants, as does the equation of a general cubic surface. There are four constants to determine the transversal, five more to fix the points where the other lines meet this, and two more to determine each of these lines, say, by fixing the point where this meets an arbitrary plane. The coefficients in the equation of the cubic surface are then functions of the constants determining the figure; and we may expect that the latter can be chosen to correspond to arbitrary values of the former. But, for this to be so, the nineteen constants of the figure must enter as nineteen effectively independent constants in the expressions, in terms of them, of the coefficients in the equation of the cubic surface. For this it is sufficient that no change in the figure of the six lines, of arbitrarily limited magnitude, shall be possible without a change in the cubic surface.

We may approach the algebraical theorem thus suggested, that a cubic surface expressed by the vanishing of a general cubic polynomial contains twenty-seven lines, in a somewhat different way. There are four algebraic conditions to be satisfied for a line to lie entirely on the surface; and a line depends on four constants. Let us assume thus that the cubic surface contains at least one line; it is then easy to infer the existence of the other lines. Taking the one line to be given by $x=0$, $y=0$, a general plane through this, of equation $x+\lambda y=0$, will meet the cubic surface further in a conic; it is at once proved that the condition for this to break up into two lines is a quintic equation for λ. By the solution of this equation we then have ten further lines on the surface, and the proof of the remaining sixteen, and that there are no more, can be completed. Or, we may initiate the investigation by seeking a plane which meets the cubic surface in three lines; the elimination of one coordinate between the equation of the cubic surface and the equation of the plane leads to a cubic polynomial in the three remaining coordinates; the three coefficients in the equation of the plane are to be chosen so that this cubic polynomial breaks up into three linear factors. In such way it can be proved that for a cubic surface whose equation has real coefficients there are at least three lines whose equations have real coefficients.

It has been shewn that the twenty-seven lines can be constructed from the figure consisting of a skew hexagon together with a transversal of three alternate sides of this hexagon. This figure, we easily see, is also of amplitude nineteen, depending on nineteen constants.

Another figure depending on nineteen constants is that of four lines of which no two intersect together with a point of general position. For a cubic surface to contain this figure requires however only seventeen conditions for the coefficients in the equation of the surface.

Ex. 4. The proof of the reciprocity of the double-six can be obtained also as follows: Consider the figure consisting of three non-intersecting lines, say a, b, c, together with three other non-intersecting lines, say d', e', f', of which every one meets all of the first three. It can be proved that this figure is obtainable, by incidences only, from five suitably chosen points of the figure, or, also, from five suitably chosen planes of the figure. Hence it can be shewn that any two such figures can be regarded as related to one another (in the sense explained in Vol. I, p. 150) so that to every point of one corresponds a point of the other; or can be regarded also as related to one another in such a way that to every point of one figure corresponds a plane of the other figure (v. Staudt, *Beiträge z. Geom. d. Lage*, I, p. 6). When the two figures are formed by the lines (a, b, c, d', e', f') and (a', b', c', d, e, f), of a double-six, it can be shewn that they can be related in the latter of the two ways referred to, with the further limitation that the relation is involutory, a point of either figure being the pole in regard to a quadric surface of the corresponding plane of the other figure. (See E. Stenfors, *Die Schläflische Konfiguration von zwölf Geraden*, Helsingfors, 1921.)

Ex. 5. An interesting special double-six has been discussed by Dr W. Burnside (*Proc. Camb. Phil. Soc.*, XVI, 1911, p. 418). Let A, B, C, D, E be five general points, taken in a definite order, forming a skew pentagon in three dimensions. From each of these points let the transversal be drawn to meet the two lines constituted by, (a), the opposite edge of the pentagon, and, (b), the join of the two vertices of the pentagon which are contiguous to that vertex from which the transversal is drawn. It can be shewn that the five transversals so obtained are all met by two lines. It can further be shewn that if the five original points be taken in all possible orders, just six such pairs of lines, each meeting five transversals described as above, are obtained. And, then, that these six pairs of lines constitute a double-six. This double-six then has the property of being unaltered by a correspondence of the three-fold space to itself, of the kind described Vol. I, p. 150, determined by the fact that the five fundamental points correspond to these same points taken in a different order. It is essentially the only possible double-six unaltered by such a (1, 1) point correspondence, or *collineation*.

The fact that the five transversals spoken of, drawn one from each of the five given points, are all met by two lines, is deducible from the theorem (above, page 68, Ex. 26) that there exists a focal system in which each of the five given points is the pole of the plane containing this and the two contiguous vertices of the pentagon; for, by this theorem, there exists also another focal system in which each point is the pole of the plane joining this point to the opposite edge of the pentagon. Analytically, if the five points be represented by the symbols A, B, C, D, E, subject to

$$A+B+C+D+E=0,$$

and m, m' be the two real numbers for which $m+m'=1$ and $mm'=-1$, it will easily be found that the thirty points obtainable by all symbols of the form $mA+B+C$, which, in another order, are the same as those obtainable by the symbols $m'A+B+C$, lie in fives upon twelve lines forming the double-six in question.

If we also take the ten points whose symbols are the differences, such as $A-B$, of the original symbols, these can be regarded, in ten ways (Vol. II, p. 212), as consisting of two triads in perspective, together with their centre of perspective and three points lying on their axis of perspective. By joining each of these ten points, regarded as centre of perspective, to the three points lying on the corresponding axis of perspective, we obtain in all fifteen new lines. These are in fact the lines c_{rs}, of our preceding theory, to be associated with the double-six above obtained.

If we take four of these ten points, which do not lie in a plane, as fundamental points, X, Y, Z, T, defined by

$$5X=A-E,\quad 5Y=B-E,\quad 5Z=C-E,\quad 5T=D-E,$$

the original five points may be defined in terms of these, namely by

$$A=4X-Y-Z-T,\ \ldots,\ E+X+Y+Z+T=0.$$

With coordinates, x, y, z, t, referred to X, Y, Z, T, the twelve lines of the double-six may be shewn to be, respectively, those given by the symbols

$$a\,(y, z, x;\ z, y, t),\quad b\,(z, x, y;\ x, z, t),\quad c\,(x, y, z;\ y, x, t),$$
$$d\,(t, x, z;\ x, t, y),\quad e\,(t, y, x;\ y, t, z),\quad f(t, z, y;\ z, t, x);$$
$$a'\,(t, x, y;\ x, t, z),\quad b'\,(t, y, z;\ y, t, x),\quad c'\,(t, z, x;\ z, t, y),$$
$$d'\,(y, z, t;\ z, y, x),\quad e'\,(z, x, t;\ x, z, y),\quad f'\,(x, y, t;\ y, x, z),$$

where, for example, the symbol $a\,(y, z, x;\ z, y, t)$ means the line given by the equations $my+z+x=0$, $mz+y+t=0$. If the first row of these be denoted by $1, 2, \ldots, 6$, and the second row by $1', 2', \ldots, 6'$, respectively, the other fifteen lines are then given, respectively, by

$$c_{35}\,(x, y+z);\ c_{26}\,(t, y+z);\ c_{34}\,(t, z+x);\ c_{15}\,(t, x+y);\ c_{14}\,(y+z, x+t),$$
$$c_{16}\,(y, z+x);\ c_{45}\,(y, z+t);\ c_{56}\,(z, x+t);\ c_{46}\,(x, y+t);\ c_{25}\,(z+x, y+t),$$
$$c_{24}\,(z, x+y);\ c_{13}\,(z, y+t);\ c_{12}\,(x, z+t);\ c_{23}\,(y, x+t);\ c_{36}\,(x+y, z+t),$$

where, for example, the symbol $c_{35}(x, y+z)$ means the line given by

$$x=0, \quad y+z=0.$$

It is easy, now, to verify that these twenty-seven lines all lie on the locus given by the equation

$$x^3+y^3+z^3+t^3-(x+y+z+t)^3=0,$$

and that the lines, 1′, ..., 6′, of the second row of the double-six, are, respectively, the polar lines of the lines, 1, ..., 6, of the first row of the double-six, in regard to the quadric whose equation is

$$x^2+y^2+z^2+t^2+(x+y+z+t)^2=0.$$

It may also be verified that one form of the equation of the cubic surface is

$$(my+z+t)(mz+x+t)(mx+y+t)$$
$$=(m'y+z+t)(m'z+x+t)(m'x+y+t).$$

The system of twenty-seven lines here considered is not the most general possible. The lines of the double-six meet in pairs in six points lying on a plane in ten different ways, these being the planes containing the sets of three of the five points A, B, C, D, E; the other fifteen lines meet in threes in ten points, as we have seen, the three concurrent lines lying in a plane. The cubic surface involved is that called by Clebsch the *diagonal* surface.

Geometrical definitions of a cubic surface. Suppose we are given any four lines, a, b, c, d, of which no two intersect, and a point, O, of general position. Let the lines be considered as two couples, a, b and c, d. If now a plane be drawn through O, and its intersections with the lines a, b be joined by a line, and its intersections with c, d be joined by another line, these two joining lines will intersect, in a point, Q, of the plane. We may consider the locus of this point Q for all possible planes drawn through the point O. The locus is in fact one of which there are three points upon an arbitrary line, l. For, consider the two quadrics (a, b, l) and (c, d, l); as they have a common generator, their common tangent planes, other than those containing the line l, form a cubic developable, of which, then, three planes pass through O. Let one of these three planes meet the line l in the point Q; we prove that this is a point of the locus spoken of. The plane in question, touching the quadric (a, b, l), contains a generator of this which meets the lines a, b, l, and, hence, passes through the point Q where the plane meets l. Similarly, a transversal of the lines c, d can be drawn, in this plane, passing through Q. Thus Q is a point of the locus. We thus obtain three points of the locus upon l, and, conversely, can shew that there are no others. When the line l, in place of being general, passes through the point O, two of the planes of the developable through O contain the line l; but there are still two points, Q, upon l, beside O, each an intersection of a line meeting a, b with a line meeting c, d. We may, thus, speak of the locus of Q as a cubic surface. Evidently this surface contains the four lines a, b, c, d; for, for instance, from any point of the line a, a transversal can be drawn to c and d, and the plane through O which contains this transversal will meet b in a point which may be joined to the point already taken on a. Also, the surface contains the transversal drawn from O to a and b; for, from any point of this, a line can be drawn to meet c and d, through which, and the former transversal, there passes a plane containing O. By a similar argument, the surface contains the transversal drawn from O to c and d.

Ex. The figure consisting of four non-intersecting lines and a point, depends on nineteen constants. If the four lines be divided into two pairs, and the two transversals be drawn from the point, one to each of these pairs, we have a new figure, depending on the same constants. And there are three such, corresponding to the three ways in which the four lines can be divided into two pairs.

The conditions that a homogeneous cubic polynomial in the four space coordinates should vanish for every point of each of the four given lines, and also for the given point, are only seventeen in number; but if the cubic polynomial is further to vanish at every point of either of the two transversals drawn from the given point to two pairs of the four given lines, two further conditions are necessary, and only one such cubic polynomial is possible. We expect then that upon the locus represented by the vanishing of such a cubic polynomial, the figure of four lines and two transversals of two pairs of these, will be obtained. We shall in fact identify this figure, below, as arising from the figure of twenty-seven lines previously discussed.

Next, suppose we are given five non-intersecting lines, a, b, c, d, e, which have a common transversal; this transversal may, for the present, be denoted by l; later it will be called f'. Then, the five lines being regarded as consisting of the two couples a, b and c, d, beside the line e, consider the two quadrics (a, b, e) and (c, d, e). As these have in common the two intersecting generators l and e, their common points, other than those on these generators, lie on a plane, and, therefore, on a conic; and their common tangent planes, other than those containing these generators, pass through a point, and, therefore, envelop a quadric cone. Let O be the vertex of this cone. Evidently the planes joining O to the six lines a, b, c, d, e, l are planes of this cone; further, the plane of the two transversals drawn from any point P, of the line e, one to the pair a, b, the other to the pair c, d, is a common tangent plane of the two quadrics, and passes through O; while, conversely, any tangent plane of the cone meets e in a definite point, P.

Now let any line, say m, be drawn through O. The two quadrics, (a, b, m) and (c, d, m), having common the generator m, meet also in a cubic curve, of which the lines a, b, c, d, as well as m, are chords. From any point, Q, of this curve, can be drawn a generator of the quadric (a, b, m), meeting the lines a, b, m, and also a generator of the quadric (c, d, m), meeting the lines c, d, m; thus Q can be constructed by drawing a plane through the line m, and, therein, joining the points where this plane meets a and b, and also the points where this plane meets c and d, and taking the intersection of these joins; the locus of such points, Q, for all such planes through m, is the cubic curve in question. As, however, the line m passes through the vertex, O, of the quadric cone above found, there are two planes through m which touch this cone; and, in each of these, by what has been said, the point Q, so obtained, is a point, P, of the line e. The cubic curve which arises from the line m has, therefore, beside a, b, c, d, m, also the line e as chord.

We thus obtain an infinite number of cubic curves having the given lines a, b, c, d, e as chords, each having as chord also a particular line, m, drawn through a point O, this point being constructed from a, b, c, d, e by a certain procedure. The aggregate of the ∞^2 points, Q, of all these cubic curves is evidently that found by drawing an arbitrary plane through O and, therein, finding the intersection of the line, joining the points where a, b meet the plane, and the line joining the points where c, d meet the plane. It is thus a cubic surface, Ω, containing a, b, c, d, the point O, and the transversals drawn from O, to meet a, b, and to meet c, d. The cubic curves obtained, having a, b, c, d and e as chords, all lie on this cubic surface, Ω; moreover, since, in any plane through O, an infinite number of lines, m, can be drawn through O, there will be an infinite number of these cubic curves passing through the point, Q, determined as above by this plane. Also, the surface Ω contains the line e, because, as we have shewn, through every point, P, of this line, there passes a plane, also containing O, for which P satisfies the condition by which points Q are determined. The surface also contains the line l, every point of which satisfies the condition of being the point Q for the plane joining O to this line.

Every four of the five lines a, b, c, d, e has, beside l, another transversal; we prove now that all these lie on the surface Ω. For the transversal, e', of a, b, c, d, this is clear, all points of this line satisfying the condition for the point Q of the plane (O, e'). Consider the transversal, d', of a, b, c, e: from the point, P, where this meets e, draw the transversal to c and d, meeting d in D; then, the join of any point, Q, of d', to D, meets c as well as d, and the point

Q lies on the line d' which meets a and b. Thus, this point Q satisfies the condition for the point Q of the plane (D, d'); and this plane passes through O because it contains the transversals from P to a, b and to c, d; by the construction made, this plane contains the line c. Denoting, similarly, by a', b', c' the transversals, beside l, respectively of (b, c, d, e), (c, a, d, e), (a, b, d, e), the point O is thus incidentally shewn to lie on the four planes $[a, b']$, $[a', b]$, $[c, d']$, $[c', d]$, which, therefore, meet in a point; these planes are in fact all tangent planes of both the quadrics (a, b, e) and (c, d, e). The line of intersection of the planes $[a, b']$, $[a', b]$, which we may denote by c_{12}, is also on the surface Ω, being the transversal drawn from O to a and b; so is the line of intersection, c_{34} say, of the planes $[c, d']$, $[c', d]$. Thus O may be defined, when c_{12}, c_{34} are given, as the intersection of these lines.

We have constructed the point O, and the cubic surface Ω, from the four lines a, b, c, d, a transversal, l, of these, and a further line, e, which meets l. Conversely, suppose a, b, c, d, and the point O, to be given, and l to be one of the transversals of a, b, c, d. Then the quadric cone, having O for vertex, which touches the five planes joining O to a, b, c, d, l, is definite; thence the quadric surface having this cone as enveloping cone and containing the two lines a, b as generators is also definite (just as, dually, the quadric containing a given conic and also two non-intersecting lines each meeting the conic once, is definite); so also, the quadric having the specified cone as enveloping cone and containing the two lines c, d as generators, is definite. Of both these quadrics the line l is a generator; these quadrics will then have common another generator, meeting l. This is the line e, which is thus constructed when a, b, c, d and O are given. However, the lines a, b, c, d have, beside l, another common transversal, already named, e'. With this, and a, b, c, d, O, we then find a further line, say f, which meets e'. As then we proceeded from a, b, c, d, e, with the common transversal l, to find O, and the cubic surface, so we can construct O, and the surface, from a, b, c, d, f, with the common transversal e'; it is easy to see that the same surface is found, and that the lines a', b', c', d', before defined, as meeting, respectively, (b, c, d, e), (c, a, d, e), (a, b, d, e), (a, b, c, e), all meet f. Thus a', b', c', d', e' have a common transversal, and, incidentally, the theorem of the double-six is obtained.

In the construction of O, and the cubic surface, from the lines a, b, c, d, e, and their common transversal, l, we have proceeded in an unsymmetrical manner in two respects, first in separating e from a, b, c, d, and then in taking the two pairs a, b and c, d from these. There are in all $5 . 3$, or fifteen, ways in which we may proceed; these lead to fifteen points O, but the cubic surface finally obtained is the same in all cases, there being only one such surface which contains the six lines a, b, c, d, e and their common transversal l, when these are general. For these involve the existence of fifteen other lines on the surface, which can be constructed from these lines alone, and the common points, of two cubic surfaces, lying in a general plane cannot be more than nine in number. If, now denoting the line l by f', we associate the lines $a, b, \ldots, e, f$; $a', b', \ldots, e', f'$, respectively, with the numbers $1, 2, \ldots, 5, 6$; $1', 2', \ldots, 5', 6'$, and denote the line of intersection of the planes $[1, 2']$, $[1', 2]$ by c_{12}, and so on, the fifteen points O will be the various points of intersection (c_{rs}, c_{ij}), where r, s, i, j are four different numbers from $1, 2, \ldots, 6$.

Ex. 1. Considering any plane through the line c_{12}, drawn to contain two further intersecting lines of the cubic surface, and any plane through the line c_{34}, drawn to contain two further intersecting lines of the cubic surface, it is the fact that either of the two former lines, meeting c_{12}, is met by one of the two latter lines, meeting c_{34}. For, otherwise, the line of intersection of these two planes, already meeting the cubic surface in the point (c_{12}, c_{34}), would meet the surface in more than three points. For instance, the plane drawn through c_{12} may contain the lines $1'$, 2, and the plane drawn through c_{34} may contain the lines 3, $4'$; then the lines $1'$, 3 meet, as also do the lines 2, $4'$. Any line of the surface meeting c_{12}, other than the lines c_{34}, c_{56}, say, for instance, the line 2, is thus met by four of the lines, other than c_{12} and c_{56}, which meet c_{34}; and the other line, $1'$, lying in the plane $[2, c_{12}]$, is met by the other four lines, beside c_{12} and c_{56}, which meet c_{34}; further, these two ranges of each four points, respectively on the lines 2 and $1'$, are in perspective from the point (c_{12}, c_{34}).

Prove that there are in all 1080 such pairs of ranges of four points of intersection, in perspective from a point of intersection of two lines of the cubic surface.

Ex. 2. Prove further that, upon the line containing the three points (c_{12}, c_{34}), $(1', 3)$, $(2, 4')$, there is a point at which intersect the planes containing, respectively, the three sets of lines $(2, 1', c_{12})$, $(4', 3, c_{34})$, (c_{24}, c_{13}, c_{56}). These nine lines, arranged in the three sets $(2, 4', c_{24})$, $(1', 3, c_{13})$, (c_{12}, c_{34}, c_{56}), lie in three other planes, making with the three former planes a pair of Steiner trihedrals. Shew also that of such lines as that containing the three intersections (c_{12}, c_{34}) $(1', 3)$, $(2, 4')$, each containing three intersections, there are, in all, 720.

Ex. 3. Prove that there exist 120 pairs of Steiner trihedrals.

The definition of a cubic surface in general. The preceding geometrical definition of a cubic surface, which has led us to the figure of twenty-seven lines previously discussed, forms a sufficient basis for the development of all the properties. Greater freedom of argument will however be available in some cases if it may be assumed that the surface so obtained has the generality of the locus expressed by the vanishing of a complete homogeneous cubic polynomial in the four coordinates of a point. We have given reasons for expecting this to be so; in what follows we shall assume that it is so. But it may be well to recognise explicitly that in so doing we omit an algebraical discussion which should logically be given.

The cubic curves of the surface which belong to a particular double-six. In the preceding discussion we have obtained a system of cubic curves lying on the cubic surface, each the intersection of two quadrics (a, b, m), (c, d, m), where m is an arbitrary line passing through the point (c_{12}, c_{34}), which was called O. Through an arbitrary point, Q, of the surface, there pass, we have seen, infinitely many such curves, these corresponding to lines, m, lying in the plane through O which contains the transversals from Q, respectively to a, b and to c, d. Through two points, Q and Q', of the surface, each of which determines such a plane, there passes one such cubic curve, corresponding to the line of intersection of these two planes. Thus two of these cubic curves intersect in one point only.

The line c_{12}, meeting a and b, and meeting the line m at O, is a generator of the quadric (a, b, m); thus, through the lines a, b, c_{12}, and any one of the cubic curves, there can be described a quadric surface. Conversely, any quadric containing a, b, c_{12} has, at O, a generator, m, of the same system as a, b, and this quadric therefore meets the cubic surface again in one of the cubic curves, which, again, lies on the quadric (c, d, m). It is clear, if we employ the algebraic symbols, that the points common to a quadric surface and a cubic surface lie on a sextic curve, that is, a curve meeting an arbitrary plane in six points; for the quadric meets the plane in a conic, of which the points are expressible by quadratic functions of a parameter, and the cubic surface meets the plane in a cubic curve, from the equation of which, by substitution of quadratic polynomials for the coordinates, a sextic equation results. Thus, when a quadric surface is drawn through the three lines a, b, c_{12}, the cubic curve in which this quadric further meets the cubic surface constitutes the whole of the residual intersection. By a similar argument, a quadric surface can be drawn through this cubic curve whose residual intersection with the cubic surface consists of the lines c, d, c_{34}. From this, by the symmetry between the lines of the cubic surface, we infer that, if r and s denote any particular two of the six lines $a, b, \ldots, e, f$, the cubic curves which we have obtained on the cubic surface are all obtainable as the residual intersections with the surface of the quadric surfaces, drawn through the three lines r, s, c_{rs}. Thence it appears, from what has been said (p. 162 above), that we are to regard the sets of three lines such as r', s', $c_{r's'}$, consisting of any two of the lines a', b', ..., e', f', and their associated line $c_{r's'}$, as particular forms of the cubic curves under discussion. But then, as the lines a', b', ..., f' are quite similar to the lines a, b, ..., f, it follows that there exists on the cubic surface another system of cubic curves, each the residual intersection with the surface of a quadric drawn, say, through the three lines $1'$, $2'$, c_{12}.

This new system of cubic curves is, however, obtainable from the former system; for it is the case that any cubic curve of the former system and any cubic curve of the latter system are, together, the complete intersection of the cubic surface with a quadric. To prove this, it is sufficient, after what has been said, to prove that, if we take the cubic curves γ and γ', of which γ is the residual intersection with the cubic surface of any quadric drawn through the lines $1'$, $2'$, c_{12}, and γ' is the residual intersection with the cubic surface of any quadric drawn through the lines 1, 2, c_{12}, then the curves γ, γ' lie on a quadric surface. This will follow if we shew that γ, γ' have five points in common, other than those on the specified lines; for, then, the quadric surface described to contain these five points, and also two other points of each of the curves, γ, γ', as then meeting each of the curves in seven points, will contain both these curves. Now, the common points of γ, γ' are the points of the cubic surface which lie on the common curve of the two general quadrics described as above, respectively through 1, 2, c_{12} and through $1'$, $2'$, c_{12}. These quadrics, having the line c_{12} in common, meet otherwise in a cubic curve; this curve passes through the two points $(1, 2')$, $(1', 2)$, which are on both quadrics, and also meets the common generator c_{12} in two points; beside the four points given by these, which lie on the cubic surface, the cubic curve will meet the cubic surface in five other points, not lying, when the quadrics drawn through the lines 1, 2, c_{12} and $1'$, $2'$, c_{12} are not special, on any of these lines. This proves the result enunciated.

From this we see that, given a single cubic curve, γ, of either system, the curves of the other system are obtainable by drawing the most general quadric surface through the curve γ. Such a quadric is identified by two points of the cubic surface not lying on γ, which also, we have seen, identify a cubic curve of the other system.

It follows now, as in a previous section (above, p. 162), that the quadric surfaces whose intersection with the cubic surface breaks up into cubic curves, one of each system, are, when expressed in point coordinates, linearly deducible from nine of them, which may be taken to be the same nine as those taken in the section referred to. These quadrics are then all outpolar to the same quadric, Σ, that, namely, in regard to which the six pairs of lines, $(1, 1'), \ldots, (6, 6')$, are polar lines of one another; and, by what we have seen, this is the same statement as that any quadric containing any cubic curve of either system is outpolar to Σ.

A particular consequence is that the five points above shewn to be common to two of the cubic curves, γ and γ', of different systems, are such that the plane which contains any three of them has, for its pole in regard to Σ, a point lying on the join of the other two points; or, in a phraseology previously employed (above, p. 37), that these five points are a self-conjugate pentad in regard to Σ. To prove this, consider the general quadric surface passing through these five points; it is a linear aggregate of five such quadrics, when expressed by its equation in point-coordinates; these may then be taken to consist of two quadrics through each of the two cubic curves which intersect in the five points, together with the quadric which contains both these curves, there being, as we know, three linearly independent quadrics through any cubic curve in space. We thus infer that every quadric through the five points is outpolar to the quadric Σ. This leads to the statement made, since a particular quadric through the five points is the plane-pair consisting of the plane through three of these points taken with any plane whatever through the other two.

Ex. 1. For the case of two degenerate cubic curves of the same system constituted, respectively, by the lines a, b, c_{12}, and by the lines c, d, c_{12}, we saw that, if a plane be drawn through the point O, or (c_{12}, c_{34}), which is common to these degenerate cubics, and the join of the two points where this plane meets the lines a, b be taken, as also the join of the two points where this plane meets the lines c, d, then these two joins meet in a point of the cubic surface, all points of the surface being obtainable by varying the plane drawn through O. We consider now a theorem which is a generalisation of this result: If any two cubic curves in space, of the cubic surface, belonging to the same system, be taken, intersecting in a point, Q, of the surface, and a variable plane be drawn through Q, meeting one of these curves again, say, in M and N, and the other in M' and N', then the lines MN, $M'N'$ meet in a point of the cubic surface, all points of which are thus obtainable by varying the plane. We base this theorem upon a property of the *plane* cubic curve in which the cubic surface meets an arbitrary plane, of which we give the proof immediately (Ex. 2); this property is that, if A, B, C, Q be any four fixed points of this plane cubic curve, and any conic be drawn through these points, meeting the plane cubic again in the two points M and N, then the third point in which the line MN meets the plane cubic again is the same whatever conic be taken. Assuming this result, let ω denote, in the theorem enunciated for the cubic curves in space on the cubic surface, the plane cubic curve in which the cubic surface is met by the plane which is drawn through Q; and let A, B, C be the points in which this plane curve ω is met by the lines a, b, c_{12}, of the cubic surface. The quadric surface containing these three lines, and also one of the two cubic curves considered, say γ, which pass through Q, will meet the plane in a conic containing the points Q, A, B, C, and also containing the two points, M and N, in which the plane meets the curve γ, beside Q. All these points are on the curve ω. Let K be the third intersection of the line MN with the curve ω. By the property of a plane cubic curve which we have assumed, this point K lies also on the line $M'N'$. Conversely, through any point, K, of the cubic surface, a single chord can be drawn of any assigned cubic curve in space, lying on the cubic surface, which passes through Q. This chord and the line KQ determine a plane. The theorem enunciated is therefore proved.

Ex. 2. Of the property, for a plane cubic curve, assumed in Ex. 1, which is a very particular case of a general theorem for plane curves, we may here give the following proof. The plane curve which is the section of the cubic surface by a plane, is met by an arbitrary line of that plane in three points; it can thus be represented by the vanishing of a homogeneous cubic polynomial in the three coordinates of a point on that plane. We have assumed that the cubic surface is that given by the vanishing of a general cubic polynomial in the four space coordinates; the plane curve will, therefore, if the plane be general, be expressed by the vanishing of a quite general cubic polynomial in the three coordinates, containing, homogeneously, ten independent coefficients. Let this equation be $f=0$, and let U, V be two other cubic polynomials in the three coordinates which both vanish at eight points of the curve $f=0$. The equation $U+\lambda V=0$ then represents a cubic curve, in the plane, passing through these eight points. We assume that the common points of the curves $U=0$, $V=0$ are nine in number, an assumption, moreover, easily verified in the application of the theorem made below. These curves will, therefore, intersect in a further point beside the given eight points, and the curve $U+\lambda V=0$ will pass through this for all values of λ. The value of λ is however particularised when a point, other than the specified ninth point, is assigned upon the curve $U+\lambda V=0$, and this curve is then uniquely determined; in particular, by taking this assigned point to lie on $f=0$, we see that the curve $f=0$ can be represented in the form $U+\lambda V=0$, if λ be properly chosen. (Cf. Ex. 6, p. 156.)

Now, suppose that through the four points A, B, C, Q, of the cubic curve $f=0$, two conics $S_1=0$, $S_2=0$ are drawn, meeting $f=0$ again, respectively, in the points M_1, N_1 and the points

M_2, N_2; let $P_1 = 0$, $P_2 = 0$ be the respective equations of the lines M_1N_1 and M_2N_2. Then the cubic polynomials S_1P_2 and S_2P_1 both vanish at the eight points A, B, C, Q, M_1, N_1, M_2, N_2 of the curve $f = 0$. Thus, by what has been said, there is an identity of the form $f = \mu S_1 P_2 + \nu S_2 P_1$, in which μ, ν are constants. This shews, however, that the lines $P_1 = 0$, $P_2 = 0$ meet on the curve $f = 0$, which proves the property in question.

The general property, of which the above is a particular case, may be enunciated for any plane curve, $F = 0$, which has no multiple points, as follows: let $A, B, C, \ldots, T$ be any set of points on $F = 0$; let $S_1 = 0$, $S_2 = 0$ be any two curves, in this plane, of the same order, both of which contain this set of points and meet the curve $F = 0$, respectively, in the further points $M_1, N_1, \ldots, U_1$ and $M_2, N_2, \ldots, U_2$, the number of points in these two further sets being in fact the same. Now, let $P_1 = 0$ be any curve whatever drawn through the points $M_1, N_1, \ldots, U_1$, having with the curve $F = 0$ the further intersections $H, K, \ldots$. The property then is that there exists a curve, of the same order as $P_1 = 0$, whose complete intersection with $F = 0$ consists of the two sets of points

$$M_2, N_2, \ldots, U_2, \ H, K, \ldots.$$

Each of the two sets $(M_1, N_1, \ldots, U_1)$ and $(M_2, N_2, \ldots, U_2)$ is said to be a *residual* of the first set, $(A, B, C, \ldots, T)$, and these two sets are therefore said to be *coresidual*. The set $(H, K, \ldots)$ is then a residual of the set $(M_1, N_1, \ldots, U_1)$. The property may then be stated by saying that any residual of one of two coresidual sets is equally a residual of the other.

Ex. 3. The property for plane curves just explained has an analogue for surfaces, in which a set of points of the plane curve is replaced by a curve on the surface, and residual sets of points by residual curves on the fundamental surface, that is, curves which, together, are the complete intersection of this surface with a further surface. We first illustrate this by considering a particular case for curves on the cubic surface. We denote the equation of the cubic surface by $F = 0$. Through any line of this surface let two planes be drawn, $S_1 = 0$, $S_2 = 0$; the residual intersections, in general, conics, may be denoted by M_1, M_2, respectively. Let an arbitrary quadric, say $P_1 = 0$, be drawn through the conic M_1, and the residual intersection of this quadric with the cubic surface be denoted by H. Then it is the case that this curve H forms, with the conic M_2, the complete intersection of the cubic surface with another quadric, say $P_2 = 0$. In general terms, the curve H, which is a residual of one of the two coresidual conics M_1, M_2, is equally a residual of the other.

To prove this, remark, first, that the curve H is a quartic curve, meeting an arbitrary plane in four points; for the aggregate of this curve and the conic M_1, being the intersection of a quadric surface and a cubic surface, meets an arbitrary plane in six points. Next, this quartic curve, H, meets every generator of the quadric $P_1 = 0$ in two points; these, with the point where the generator meets the conic M_1, are the three intersections of the line with the cubic surface, when this does not contain the line entirely. A quartic curve, on a quadric, meeting every generator in two points, is however the complete intersection of this quadric with another quadric, and a quadric can be drawn through the curve and an arbitrary point. Assuming this fact, which is proved below, consider the four intersections of the quartic curve with the plane of the conic M_2; as the curve is on the cubic surface these intersections, if not all on the conic M_2, can only be on the line of intersection of the planes of the two conics M_1 and M_2, and, therefore, as the curve is on the quadric $P_1 = 0$, can only be where the conic M_1 meets this line, which requires a particular quadric P_1. Thus, in general, the curve will meet the conic M_2 in four points. Therefore a quadric can be drawn through the quartic curve to contain also the conic M_2; for, we have only to describe a quadric through the quartic curve which shall meet this conic in a fifth point other than those four in which the quartic curve already meets the conic.

To prove the fact stated, that if a quartic curve, H, lying on a quadric surface, P, meet every generator of the quadric in two points, then a quadric can be drawn through the curve and any other arbitrary point, we may argue as follows: Let the arbitrary point be O, and take eight points, $A_1, A_2, \ldots, A_8$, of general position, upon the quartic curve; describe the quadric surface, P', through O and these eight points; this will meet the quadric P in a quartic curve, H'. We prove that this coincides with the curve H. For, first, this curve H' meets every generator of the quadric P in two points, as we see by considering how many points of this curve, not on this generator, lie on an arbitrary plane drawn through this; this plane meets the quadric P, further, in a line, and meets the quadric P' in a conic. Wherefore, if we project the two curves H, H', on to an arbitrary plane, from one of the eight points, $A_1, \ldots, A_8$, which are common to both curves, say from A_1, we obtain two plane cubic curves having nine points in common, namely the projections of the seven points $A_2, \ldots, A_8$, and the points where the generators of the quadric P, at the point A_1, meet this plane. We have proved above (Ex. 2) that, of the nine intersections of a given plane cubic curve (here the projection of the curve H) with another such curve, only eight can be taken arbitrarily upon the first curve, the remaining one being then determinate. In our present case, by retaining the same positions for $A_1, A_2, \ldots, A_7$, and changing the position of A_8, we can, if the curve

H' is distinct from H, assume the ninth intersection of two such plane cubic curves quite arbitrarily upon one of them. We infer therefore that H' coincides with H. The fact stated is then proved.

Ex. 4. The general property referred to, at the beginning of Ex. 3, may be stated for a cubic surface (to which however it is not restricted) as follows: Suppose that, through any curve, l, on the cubic surface, there pass two surfaces of equal order, $S_1 = 0$, $S_2 = 0$, whose residual intersections with the cubic surface are, respectively, the curves M_1 and M_2; let any surface whatever, $P_1 = 0$, be drawn through the curve M_1, having, as its residual intersection with the cubic surface, a curve H; then a surface, of the same order, exists, whose complete intersection with the cubic surface consists of the curves H and M_2. It seems desirable to enunciate this property; but we give only indications of the proof. This depends on the fact that, if $F = 0$ be the equation of the cubic surface, and $S = 0$ the equation of another surface, then any surface containing the complete intersection of $F = 0$ and $S = 0$ has an equation of the form $\Theta F + PS = 0$; here Θ, P are homogeneous polynomials in the coordinates, of respective orders such as to make the left side of the equation of the order of the surface which is to be expressed. In our case take for S the surface $S_1 = 0$, whose intersection with the cubic surface consists of the curves l and M_1; consider the composite surface consisting of the surfaces $S_2 = 0$ and $P_1 = 0$, with equation $S_2 P_1 = 0$; as $S_2 = 0$ contains the curve l, as well as the curve M_2, and $P_1 = 0$ contains the curve M_1, as well as the curve H, this surface $S_2 P_1 = 0$ contains the intersection, (l, M_1), of the surfaces $F = 0$, $S_1 = 0$; there is thus, by what we have stated, an identity of the form $S_2 P_1 = \Theta F + P_2 S_1$, wherein Θ, P_2 are appropriate polynomials. The surface expressed by $P_2 = 0$ thus meets the cubic surface, $F = 0$, on the portion of the intersection of this with $S_2 P_1 = 0$ which does not lie on $S_1 = 0$. Wherefore, the curves M_2 and H lie on the surface $P_2 = 0$; it was the existence of such a surface which was to be proved.

The development of the properties of coresidual curves on a cubic surface is of great importance, but would carry us too far. The reader may consult a paper, *Proc. Lond. Math. Soc.* XI, 1912, p. 287.

Ex. 5. The quadrics which pass through the curve of intersection of two given quadrics meet an arbitrary line in pairs of points of an involution. From the result of Ex. 3, it follows that the conics in which the cubic surface is met by the planes drawn through a line of the surface, say l, meet this line in pairs of points of an involution. Of this result another proof arises below. The double points of the involution are two important points of the line; in fact it will be found below that the line touches at these two points the quartic surface known as the Hessian of the cubic surface.

Ex. 6. The theorem stated in Ex. 5 is easily obtained analytically; and another result which includes this as a particular case may be obtained at the same time. Let $F=0$ be the equation of the cubic surface, and x_1, y_1, z_1, t_1 be the coordinates of any particular point, P_1. The quadric surface whose equation is

$$[x_1\partial/\partial x + y_1\partial/\partial y + z_1\partial/\partial z + t_1\partial/\partial t]\, F = 0$$

is called the polar quadric of the point P_1 in regard to the cubic surface. It is such that the line joining P_1 to any point, P, of the curve of intersection of the cubic surface with this polar quadric of P_1, meets the cubic surface in two coincident points at P, as we may easily see by considering the equation in λ obtained by substituting in $F=0$, for the coordinates, the values

$$x+\lambda x_1,\ y+\lambda y_1, \text{ etc.}$$

The general result referred to is that, if we take any particular line lying on the cubic surface, the ∞^3 quadrics constituted by the polar quadrics of all points P_1 meet this line in pairs of points of an involution. That this involution includes the pairs of points of intersection, with the line, of conics lying on the cubic surface, in planes passing through the line, is clear; for a point, P, where such a conic meets the line, arises twice over in the intersections with the cubic surface of any line through P lying in the plane of the conic. To prove the result, let the coordinates be so taken that the line of the cubic surface which is considered is that given by $x=0, y=0$; the equation of the cubic surface is then of the form $xU+yV=0$, where $U=0$, $V=0$ represent quadric surfaces. The polar quadric, by the form of its equation, is unchanged by any particular choice of coordinates (cf. Vol. II, p. 104). In the present case, the equation of the polar quadric of P_1 is

$$x_1U + y_1V + xDU + yDV = 0,$$

where D represents the operator by which the polar quadric, $DF=0$, is formed. The polar quadric thus meets the line $x=0$, $y=0$ in the pair of points given by $x_1(U)+y_1(V)=0$, where (U), (V), containing terms in z^2, zt, t^2, represent the result of putting $x=0$, $y=0$ in U and V. We know that the pairs of points given by this equation, when x_1, y_1 vary, belong to an involution (Vol. II, p. 111, Ex. 3). The general plane through the line $x=0$, $y=0$, say $y-\lambda x=0$, meets the cubic surface on the quadric cone expressed by $[U]+\lambda[V]=0$, where $[U]$, $[V]$ are obtained from U, V respectively, by substituting λx for y. This cone meets $x=0, y=0$ in a pair of points given by $(U)+\lambda(V)=0$, belonging to the same involution as before.

The double points of the involution, being harmonic in regard to the intersections of the line with a polar quadric, are conjugate to one another in regard to all polar quadrics. But there exist pairs of points with this property not lying on a line of the cubic surface, as we shall see (see below, under ***Hessian surface***).

Ex. 7. A result given above suggests another way in which we may approach the theory of a cubic surface and the lines lying thereon. Suppose that we are given two cubic curves, γ and γ', in three dimensions, having a point in common, say O. An arbitrary plane, drawn through O, will meet these curves both in two further points, say, respectively, A, B, and A', B'. We may consider the locus of the point of intersection, P, of the two lines AB, $A'B'$, as the plane drawn through O varies. It follows from what has preceded that the locus is a cubic surface containing the curves γ, γ'. But this may be proved directly. For, take nine points, other than O, on each of these cubic curves. The nineteen points thus obtained suffice to determine a cubic surface, defined as the locus expressed by a cubic polynomial in the four space coordinates. This surface will contain both the cubic curves, as meeting each in ten points, nine being the number of intersections of a cubic surface with a cubic curve. There are, in general, ten common chords of two cubic curves in space, as has been seen (above, p. 141); the quadric cones projecting the cubic curves γ and γ' from O have four common generators, which are among these ten. There remain, then, six other common chords of these two cubic curves, say, a, b, c, d, e, f; and each of these chords, as meeting the cubic surface in four points, will lie entirely thereon. It can be shewn that, taking two of these common chords, say a and b, there is a single line meeting both, and also meeting each of the curves γ, γ', in such a way that the four points of meeting are distinct. This line will then also lie on the cubic surface. In general, four curves in space, of respective orders m, n, p, q, are met by $2mnpq$ lines (see below); this would give eighteen lines meeting both a and b and also meeting each of the curves γ, γ' once. We can, however, with more careful examination, see that seventeen of these lines are such as to meet the cubic surface in less than four points; it cannot, therefore, be inferred that these lie on the surface. For, let the line a meet the curves γ and γ', respectively, in the points A_1, A_2 and A_1', A_2', and the line b meet γ and γ' respectively, in B_1, B_2 and B_1', B_2'. Then, one of the seventeen lines is the transversal from O to a and b; two more are the joins of A_1 to B_1' and B_2', and there are eight such lines in all; if the quadric be described containing the curve γ' and its two chords, a, b, there will be a generator of this through A_1, of the opposite system from these, which will meet the curve γ', as well as a, b and the curve γ; this is another of the seventeen lines, and there are four such, respectively through A_1, A_2, B_1, B_2. There are four similar lines definable by the quadric containing the curve γ and its chords a, b. Seventeen lines are thus accounted for, each of which meets the cubic surface in three points only. There is, thus, as stated, one line meeting the curves γ, γ' and the lines a, b, in four distinct points, and, therefore, lying on the cubic surface. We may denote this by c_{12}. By taking every couple of the six lines $a, b, \ldots, e, f$, we obtain in all fifteen such lines of the cubic surface; the theory of the surface can then be developed from this point.

To prove that there are $2mnpq$ lines each meeting four given curves, of respective orders m, n, p, q, we may assume that a surface of order N is met by a curve in space, of order k, in Nk points. If the surface and curve are defined by algebraic equations this is an algebraic theorem, which needs proof. Assuming this, however, consider first the particular case of the result under discussion which asserts that there are $2m$ lines meeting one curve of order m and three lines. The transversals of these three lines generate a quadric surface, which, by the assumption made, meets the curve of order m in $2m$ points; from each of these there is one of the lines desired, meeting the curve of order m and the three given lines. Hence we have the result that the lines meeting a curve of order m, and also meeting two lines, generate a surface of order $2m$. By the assumption made, this surface will meet a curve of order n in $2mn$ points; this is, then, the number of lines meeting two given curves of orders m and n, and also meeting two given lines. Wherefore, the lines meeting two given curves of orders m and n, and also meeting a line, generate a surface of order $2mn$. By the assumption made, this surface will meet a curve of order p in $2mnp$ points; this is, then, the order of the surface formed by the lines meeting three given curves of respective orders m, n, p. Hence, this surface meeting a curve of order q in $2mnpq$ points, there are $2mnpq$ lines meeting four curves of respective orders m, n, p, q.

Definition of a cubic surface by three related star-figures. We have explained in Vol. I (p. 148) what is meant by two related plane systems, wherein, to any point of one plane there corresponds a point of the other, and to a range of points in one plane there corresponds a related range of points of the other. Suppose the points and lines of one of these planes are joined to a point, O, not lying in this plane, respectively by lines and planes; and, similarly, that the points and lines of the other plane are joined to a point, O', not lying in this plane, respectively by lines and planes. We then obtain two star-figures, or central-systems, with vertices, or centres, respectively at O and O', which are said to be related to one another, any line, or flat pencil of lines, passing through O, corresponding to a line, or related flat pencil of lines, passing through O', and conversely.

Such a relation between two star-figures can be set up by means of a cubic curve which contains the two vertices O and O'. It is only necessary to make correspond to any plane through O, which will meet the cubic curve in two further points, the plane joining O' to the chord of the cubic curve which contains these points; then, any line through O, which meets an infinite number of chords of the curve, whose ends are pairs of an involution thereon, will correspond to the line through O' which meets all these chords. It may be proved that any flat pencil of lines through O gives rise in this way to a related flat pencil of lines through O'; and, conversely, that any two related star-systems may be regarded as arising in this way from a cubic curve in space.

Now suppose we have three star-systems, of vertices O, O' and O'', of which any two are related to one another, so that, to any plane through O, there corresponds a definite plane through O' and a definite plane through O''. We may then consider the locus of the point of intersection of three such related planes. By what has been said, this comes to supposing that we have two cubic curves which meet in O, one of these containing O', the other containing O'', and that we draw through O an arbitrary plane, meeting the first cubic curve in the further points A', B' and the second curve in the further points A'', B'', and then consider the locus of the intersection of the lines $A'B'$ and $A''B''$. We are thus brought back to the definition of a cubic surface which has been considered (Ex. 7, p. 183); and it appears, since it has been shewn that, on a cubic surface, there can be drawn one cubic curve of a definite system through two arbitrary given points, that a cubic surface can be generated by the intersection of corresponding planes of three related star-systems in which the vertices, O, O', O'', are any three points of the surface.

By choosing a proper system of coordinates, related to a cubic curve containing the points O and O', the corresponding planes through these points, intersecting in a chord of this curve, can (above, p. 131) be supposed to have the equations

$$x - y(\theta + \phi + \omega) + z(\theta\omega + \phi\omega + \theta\phi) - t\theta\phi\omega = 0,$$
$$x - y(\theta + \phi + \omega') + z(\theta\omega' + \phi\omega' + \theta\phi) - t\theta\phi\omega' = 0,$$

or say $\xi u + \eta v + \zeta w = 0, \quad \xi u' + \eta v' + \zeta w' = 0,$

wherein $\xi : \eta : \zeta, = 1 : \theta + \phi : \theta\phi$, have ratios which vary as these planes vary, but $u, = x - y\omega$; $v, = -y + z\omega$; $w, = z - t\omega$; $u', = x - y\omega'$, etc., are fixed. The planes $u = 0$, $v = 0$, $w = 0$, of the star-figure of vertex O, correspond then, respectively, to the planes $u' = 0$, $v' = 0$, $w' = 0$, of the star-figure of vertex O'. When we take any general system of coordinates the equations of corresponding planes of the star-figures (O), (O') will remain of the forms $\xi u + \eta v + \zeta w = 0$,

$\xi u' + \eta v' + \zeta w' = 0$, and the plane of the star-figure (O'') which corresponds to these will be of the form $\xi u'' + \eta v'' + \zeta w'' = 0$, where u'', v'', w'' are fixed planes through O'', corresponding respectively to u, v, w. The equation of the cubic surface is thus

$$\begin{vmatrix} u, & v, & w \\ u', & v', & w' \\ u'', & v'', & w'' \end{vmatrix} = 0.$$

This form gives rise to several remarks. In the first place, considering the three quadrics which are represented by $vw' - v'w = 0$, $wu' - w'u = 0$, $uv' - u'v = 0$, the last two of these have in common the line given by $u = 0$, $u' = 0$, and, beside this, intersect in a cubic curve which lies on the first. Thus this cubic curve lies on the cubic surface. It is evidently the cubic curve joining the points O and O'. Similarly, the cubic curve of the surface joining the points O and O'' is that common to the three quadrics $vw'' - v''w = 0$, $wu'' - w''u = 0$, $uv'' - u''v = 0$. We may, however, without altering the equation of the cubic surface, replace u, v, w, respectively, by $au + bu' + cu''$, $av + bv' + cv''$, $aw + bw' + cw''$, where a, b, c are any constants; and at the same time make similar replacements for u', v', w', and for u'', v'', w''. Thereby we obtain an infinite number of cubic curves lying on the surface; it may be seen that these are such as those previously described as belonging to one system. Any one of these curves is obtained by taking two of the rows of the modified determinantal equation of the cubic surface; by dealing similarly with the *columns* of the determinant, another system of cubic curves of the surface is obtained. Moreover, as any minor of the determinant, of two rows and columns, belongs as well to two rows as to two columns, we see that any cubic curve of the first system and any cubic curve of the second system are together the complete intersection of the cubic surface with a quadric, as was previously proved. Further, the equations of all quadric surfaces so arising are linear functions of those nine of them which correspond to the nine minors of the determinant, of two rows and columns, as also has been shewn above.

Ex. 1. The six lines which form one row of the double-six associated with the present way of writing the equation of the cubic surface are the common chords of one system of cubic curves. To find them we are to express that the planes $\xi u + \eta v + \zeta w = 0$, $\xi u' + \eta v' + \zeta w' = 0$, $\xi u'' + \eta v'' + \zeta w'' = 0$, meet in a line; this requires an identity, in regard to the coordinates, of the form $\Sigma\xi(\lambda u + \lambda' u' + \lambda'' u'') = 0$, in which the three parameters λ, λ', λ'' are independent of the coordinates; this again, leads to four equations all linear both in ξ, η, ζ and in λ, λ', λ''. By elimination of λ, λ', λ'', from each set of three of these equations, we thence obtain cubic polynomials in ξ, η, ζ; the values of these for which these polynomials vanish solve the problem. The other row of the double-six is to be similarly found from the columns of the determinant.

Ex. 2. The nine independent quadrics, each containing a cubic curve, on the cubic surface, of each of the two systems, are outpolar to a single quadric, Σ; in regard to this quadric, as has been seen, the lines of one row of the double-six are polar lines of those of the other row, respectively. This quadric, Σ, is inpolar to any quadric which, expressed in point coordinates, is a linear function of the nine quadrics. Now, the cubic surface being $F = 0$, let $DF = 0$ be the polar quadric (above, p. 182) of any point (x_1, y_1, z_1, t_1), where the operator D is a sum of such terms as $x_1\partial/\partial x$. This is a linear function of the nine quadrics referred to; for, with the above determinant form for F,

$$DF = \begin{vmatrix} Du, & Dv, & Dw \\ u', & v', & w' \\ u'', & v'', & w'' \end{vmatrix} + \begin{vmatrix} u, & v, & w \\ Du', & Dv', & Dw' \\ u'', & v'', & w'' \end{vmatrix} + \begin{vmatrix} u, & v, & w \\ u', & v', & w' \\ Du'', & Dv'', & Dw'' \end{vmatrix},$$

and each of Du, Dv, etc., is a constant, being the values of u, v, etc. for the point (x_1, y_1, z_1, t_1); the minors, in the determinants, multiplying these constants, are the nine quadrics in question.

We infer, therefore, that the quadric Σ is inpolar to every polar quadric. The expression for F by which we have deduced this is connected with one of the double-sixes of the lines on the cubic surface, and the two systems of cubic curves associated therewith; the quadric Σ is also related to this double-six. There is such a quadric for every one of the double-sixes, and, therefore, there are in all thirty-six such Schur quadrics. All these will then be inpolar to the polar quadrics, that is, inpolar to the four quadrics

$$\partial F/\partial x = 0, \ \ldots, \ \partial F/\partial t = 0.$$

Hence, the *tangential* equations of the Schur quadrics are all linear functions of at most six of them.

We have already (above, p. 182) considered pairs of points which are conjugate to one another in regard to all the polar quadrics of the cubic surface. These form, in fact, pairs of so-called corresponding points of a certain quartic surface, the Hessian of the cubic surface. Each such pair may be regarded, tangentially, as a degenerate quadric which is inpolar to the polar quadrics; thus the tangential equations of all such point-pairs may be expressed linearly by those of six of them. The Schur quadrics, in tangential form, may then be expressed by such six point-pairs.

Ex. 3. For the cubic surface whose equation is expressed by

$$ax^3 + by^3 + cz^3 + dt^3 + eu^3 = 0,$$

where u is expressible in terms of x, y, z, t by means of the relation

$$ax + by + cz + dt + eu = 0,$$

the general polar quadric is at once seen to be expressible as a linear function of the four $x^2 - u^2 = 0$, $y^2 - u^2 = 0$, $z^2 - u^2 = 0$, $t^2 - u^2 = 0$, namely to be of the form

$$Ax^2 + By^2 + Cz^2 + Dt^2 + Eu^2 = 0,$$

with $A + B + C + D + E = 0$. Thus a pair of points, (x, y, z, t, u), $(x_1, y_1, z_1, t_1, u_1)$, which are conjugate to one another in regard to all these quadrics must be such that

$$xx_1 = yy_1 = zz_1 = tt_1 = uu_1;$$

they must, therefore, be such that both of them satisfy the equation

$$ax^{-1} + by^{-1} + cz^{-1} + dt^{-1} + eu^{-1} = 0,$$

which, conversely, is also a sufficient condition. It may easily be verified that the polar quadric of either of the two points, in regard to the cubic surface, is a quadric cone whose vertex is at the other point; this, as may easily be seen, is the same thing as that the two points lie on a surface whose equation can be expressed by the vanishing of a determinant of four rows and columns in which the elements of the first row are $\partial^2 F/\partial x^2$, $\partial^2 F/\partial x \partial y$, etc., those of the second row are $\partial^2 F/\partial y \partial x$, $\partial^2 F/\partial y^2$, etc., and so on. For the form of F under discussion, this equation reduces to that given, $ax^{-1} +$ etc. $= 0$. As has been said, the surface is called the Hessian. With this form for F, the equation of the Hessian can be seen to be capable, when u is expressed in terms of x, y, z, t, of the form

$$\lambda + bcX + adP + caY + bdQ + abZ + cdR = 0,$$

where

$X = yz^{-1} + y^{-1}z$, $P = xt^{-1} + x^{-1}t$, etc., and $\lambda = a^2 + b^2 + c^2 + d^2 - e^2$. This shews, however, that any seven points of the Hessian, of coordinates (x_r, y_r, z_r, t_r), for $r = 1, \ldots, 7$, are connected, when $A_1, \ldots, A_7$ are suitably chosen, by the equation

$$\sum_{r=1}^{7} A_r (ux_r + vy_r + wz_r + pt_r)(ux_r^{-1} + vy_r^{-1} + wz_r^{-1} + pt_r^{-1}) = 0,$$

holding for all values of u, v, w, p. This is the equation connecting the tangential equations of any seven pairs of corresponding points on the Hessian.

It may also be proved that the tangential equation of a general quadric which is inpolar to all the polar quadrics of the cubic surface is a linear function of the six quadrics such as

$$u^2 + v^2 + w^2 + p^2 - \lambda vw/bc = 0, \ldots, \; u^2 + v^2 + w^2 + p^2 - \lambda up/ad = 0,$$

where λ has the same value as above. Every Schur quadric is thus expressible by these six quadrics.

The representation of the cubic surface on a plane. In the case of curves we have distinguished between those, such as the conic and the cubic curve in space, of which the coordinates of a point are expressible by rational functions of a parameter, and those, much the larger class, for which this is not so. The examples of the latter class which we have met with most often are the cubic curve in a plane, and the quartic curve which is the intersection of two quadrics. In the case of surfaces a similar distinction is observable: there are surfaces of which the coordinates of a point are expressible as rational functions of two parameters, these parameters being themselves definable each as a rational function of the coordinates of a point of the surface; but for the majority of surfaces this is not possible; nor is it easy, as it is for curves, to give a general criterion sufficient to ensure its possibility. A quadric is a surface of which any point is given by the intersection of two generators, each of which is determined by a single parameter, and the expression in question is thence possible. We now proceed to remark on the fact that a cubic surface also has this character. In general, the two parameters, in terms of which the points of such a surface are expressible, may be interpreted as ratios of the coordinates of a point in a plane; but it is possible, in an infinite number of ways, to set up a correspondence, between any point of a plane and some other point and conversely, expressible by rational equations. Thus, when a surface is representable on a plane in one way, it is so representable in an infinite number of ways. In general, any such representation must be expected to have *singular* elements; a point of the surface may correspond, not to a point, but to a curve, of the plane; for instance, when a quadric is projected on to a plane from a point of the quadric, the point gives rise to a line of the plane; or, a curve of the surface may lead only to a point in the plane, as we shall see in the case of the cubic surface. The various representations possible for a given surface are to be distinguished by the character of these singular elements.

The generation of the cubic surface by the intersection of corresponding planes of three related star-systems leads at once to a representation of the surface upon a plane. For we may regard the planes of one, and therefore of all, of the star-systems, as being related to the points of a plane. With symbols, the cubic surface being obtained, as above, by the locus of the intersection of three planes $\xi u + \eta v + \zeta w = 0$, $\xi u' + \text{etc.} = 0$, $\xi u'' + \text{etc.} = 0$, we may represent this intersection by a point of coordinates ξ, η, ζ, in a plane. These are then proportional to quadric functions of the coordinates, x, y, z, t, of the point of the cubic surface; conversely, x, y, z, t, obtained from the three equations such as $\xi u + \text{etc.} = 0$, are proportional to cubic polynomials in ξ, η, ζ, say, respectively, $\vartheta, \phi, \psi, \chi$.

The correspondence between the planes $\xi u + \text{etc.} = 0$, $\xi u' + \text{etc.} = 0$, was obtained by the fact that these intersect in a chord of a certain cubic curve, say γ, each point of the curve being an intersection of corresponding rays of two star-systems; through each point of this curve there pass then an infinite number of pairs of corresponding planes of these two star-systems, and the cubic curve is the locus common to the three quadric surfaces $vw' - v'w = 0$, $wu' - w'u = 0$, $uv' - u'v = 0$. Another cubic curve, γ', common to $vw'' - v''w = 0$, $wu'' - w''u = 0$, $uv'' - u''v = 0$, determines the correspondence between the planes $\xi u + \text{etc.} = 0$, $\xi u'' + \text{etc.} = 0$. These curves meet in a point, O, where $u = 0$, $v = 0$, $w = 0$. From any general point, P, of the cubic surface, a chord can be drawn to each of the curves γ, γ'; the plane of these chords is the plane $\xi u + \text{etc.} = 0$, passing through the point O, and this determines the two corresponding planes of the other two systems, and the ratios $\xi : \eta : \zeta$. In particular, when P is on the curve γ, the chord of γ' drawn from P determines the plane $\xi u + \text{etc.} = 0$, and this determines the chord of the curve γ through which the plane $\xi u' + \text{etc.} = 0$ also passes. Thus, any point of the cubic surface determines a single determinate point, (ξ, η, ζ), of the plane. But, conversely, when P is on a line which is a common chord, not passing through O, of the curves γ, γ', the point (ξ, η, ζ) arises equally well from any point of this chord. We have already remarked that there are six such common chords (above, p. 183); thus there are six points of the plane (ξ, η, ζ) which correspond to lines, say a, b, c, d, e, f, of the cubic surface. This result is obtainable also by considering the sections of the cubic surface by planes which pass through a line, say l; if one such plane have the equation $Ax + By + Cz + Dt = 0$, its section with the cubic surface corresponds to the plane curve

$$A\vartheta + B\phi + C\psi + D\chi = 0,$$

which contains the three points of the plane corresponding to the intersections of the line l with the surface. A point of the plane (ξ, η, ζ), other than one of these three points, which is common to two such plane curves $A\vartheta + \text{etc.} = 0$, $A'\vartheta + \text{etc.} = 0$, corresponds, therefore, to *two* different points of the cubic surface, lying respectively on the planes $Ax + \text{etc.} = 0$, $A'x + \text{etc.} = 0$. Two plane cubic curves have, in fact, nine common points. Thus the four curves $\vartheta = 0$, $\phi = 0$, $\psi = 0$, $\chi = 0$ have six points in common; these are the six points referred to.

The loci in the plane (ξ, η, ζ) which correspond to the known loci of the cubic surface can now be enumerated. If, through the point O, there be drawn an arbitrary line, and $\xi_1 u + \eta_1 v + \zeta_1 w = 0$, $\xi_2 u + \eta_2 v + \zeta_2 w = 0$ be two planes which contain this, the corresponding lines of the other two star-systems are determined; to any plane through the first line correspond definite planes through the two other corresponding lines, and the locus of the intersection of these three planes, when the first plane varies, is a cubic curve. The points of this correspond to the points $(\xi_1 + \theta\xi_2, \eta_1 + \theta\eta_2, \zeta_1 + \theta\zeta_2)$, of the plane (ξ, η, ζ), obtained by varying θ. Such cubic curves of the cubic surface, of which, for instance, one particular case is the curve common to the three quadrics $uv' - u'v = 0$, $uv'' - u''v = 0$, $u'v'' - u''v' = 0$, thus correspond to the lines of the plane (ξ, η, ζ). The cubic curves of the cubic surface which are of the same family as the curves γ, γ', on which the vertices of the star-systems lie, meet the cubic curves just obtained, each in five points, and, moreover, meet each of the lines $a, b, \ldots, f$ in two points. The cubics of the cubic surface of the family of γ and γ' thus correspond to curves, in the plane (ξ, η, ζ), which are of the fifth order and have a double point at each of the six singular points of this plane. A particular case of a cubic of the same family as γ, γ' is the aggregate of the lines a', b', c_{12}; a line a' meets five of the lines a, b, c, d, e, f, and is a chord of cubics of the family complementary to γ, γ'. Thus the line a' corresponds to a curve, in the plane (ξ, η, ζ), which contains the five singular points corresponding to b, c, d, e, f, and meets a line of this plane in two points; namely, this curve is the conic containing the five singular points. The line c_{12}, which meets a, b and also a', b', corresponds to the join of the two singular points $(a), (b)$; the degenerate quintic curve corresponding to (a', b', c_{12}) consists of two conics and this line. The fifteen lines c_{rs} thus correspond to the joins of the six singular points.

More generally, it appears that, if we use the symbol U to denote, either an arbitrary line in the plane (ξ, η, ζ), or the corresponding curve of the cubic surface, of the family not meeting any of the lines a, b, c, d, e, f, and use the symbols $A_1, A_2, \ldots, A_6$ to denote, either the six singular points of the plane, or the corresponding lines $a, b, \ldots, f$ of the cubic surface, then a curve of order λ, in the plane, which has a general multiple point of order λ_r at the point A_r $(r = 1, \ldots, 6)$, may be associated with the symbol

$$\lambda U - \lambda_1 A_1 - \ldots - \lambda_6 A_6;$$

and that this will denote then, equally, a curve of the cubic surface of order $3\lambda - \lambda_1 - \ldots - \lambda_6$. It is necessary to prove (see below, Ex. 5), but it is stated here for clearness, that this curve is obtainable by the following process: Draw, through a set of λ cubic curves of the cubic surface, chosen arbitrarily from among those of the family to which U belongs, any surface, Ω, of sufficiently high order to contain these, and consider the curve which is the residual intersection of this surface with the cubic surface; through this residual curve draw the general surface, Ω', of the same order as Ω, which contains the line a as a λ_1-fold line, the line b as a λ_2-fold line, and so on; the remaining intersection of this second surface, Ω', with the cubic surface, is a curve which is represented by the symbol $\lambda U - \lambda_1 A_1 - \ldots - \lambda_6 A_6$; and there is a family of such curves when the conditions for the surface Ω' do not determine it uniquely. The curves of the family U are the simplest example of the statement, these being determined, as we have seen, by quadric surfaces all passing through an arbitrary cubic curve of the complementary family. The lines of the cubic surface such as c_{12} form another example; on the plane (ξ, η, ζ) this line, passing through A_1 and A_2, is representable by $U - A_1 - A_2$; on the cubic surface, we have seen, the aggregate (c_{12}, a, b) form a particular curve of the family of curves U. The representation of a conic, on the plane, passing through $A_2, A_3, \ldots, A_6$, will be $2U - s + A_1$, where s denotes $A_1 + \ldots + A_6$; it has been seen that this conic represents the line a' of the cubic surface. That this follows from the rule stated above we may see by drawing quadrics through two of the curves U, and considering a quartic surface through the residual intersection of these quadrics with the cubic surface; this residual intersection consists of two cubic curves of the complementary system; the quartic surface through these two cubic curves, and through the lines b, c, d, e, f, will further meet the cubic surface in a locus of order $3 . 4 - 2 . 3 - 5$, that is, in a line. This is then the line a', which, meeting the quartic surface in the five points in which it meets the lines b, c, d, e, f, lies entirely thereon. The representation, upon the plane (ξ, η, ζ), of a plane section of the cubic surface, being by a cubic curve through the singular points, will be associated with the symbol $3U - s$. This is the same as

$$(U - A_1 - A_2) + (2U + A_1 - s) + A_2,$$

in accordance with the fact that, on the cubic surface, the lines c_{12}, a', b form a particular plane section. We may easily interpret the symbol $3U - s$ directly, on the cubic surface, in accordance with the rule given above; for instance by using a sextic surface passing through three cubic curves of the family complementary to U.

Ex. 1. With some loss of symmetry we may obtain the representation of a cubic surface upon a plane in what is perhaps a more natural way, as follows. Assume that the surface possesses two non-intersecting lines; then, from any point, P, of the surface, a definite transversal can be drawn to these two lines, and this will meet an assumed fixed plane in a definite point, Q; conversely, from Q we can in general pass back to P without ambiguity, also by drawing a transversal, and considering its third intersection with the cubic surface. In this representation, all the points of the cubic surface which lie on a line, lying on the cubic surface, which is a transversal of the two given lines, correspond to the same point, Q, of the plane. Through either of the two assumed lines there pass five planes whose further intersection with the cubic surface consists of two lines, as has been remarked; the other of the two assumed lines thus meets at least one of each of these five pairs of coplanar lines. That is, there are five lines on the cubic surface meeting both the two assumed lines. Denote the two assumed lines by a' and b', and one of the five lines meeting both these by c_{12}; take coordinates so that the line a' is $x=0$, $y=0$, the line b' is $z=0$, $t=0$, and the line c_{12} is $y=0$, $t=0$; these leave a certain arbitrariness in x and z. It can then be shewn that the equation of the cubic surface is of the form

$$x\{z(ht-h'y)+t(gt-a'x-g'y)\}$$
$$+y\{z(bz+ft-f'y)+t(ct-c'y)\}=0.$$

Now take a point (ξ, η, ζ), in a plane, such that

$$\xi:\eta:\zeta=xt:yz:yt;$$

this point then corresponds to a definite line meeting the lines a', b', and so to a definite point of the cubic surface; and points of a line, $l\xi+m\eta+n\zeta=0$, correspond to the points on a cubic curve of the cubic surface, lying on a quadric $lxt+myz+nyt=0$, which is the general quadric containing the lines a', b', c_{12}. Conversely, we find, for the point (x, y, z, t) of the cubic surface,

$$x:y:z:t=\xi\phi:\zeta\phi:\eta\psi:\zeta\psi,$$

where $\phi=0$, $\psi=0$ represent conics in the plane (ξ, η, ζ). Thus a plane section of the cubic surface corresponds to a cubic curve, in the plane (ξ, η, ζ), with the equation

$$(A\xi+B\zeta)\phi+(C\eta+D\zeta)\psi=0.$$

This, whatever A, B, C, D may be, has six definite points: the intersections of the conics $\phi=0$, $\psi=0$; and the two points, $\xi=0$, $\zeta=0$, on $\psi=0$, and $\eta=0$, $\zeta=0$, on $\phi=0$. The representation can now be compared in detail with that given above.

Ex. 2. The symbol, for a cubic curve of the cubic surface, of the family complementary to U, represented on the plane by a quintic curve having a double point at each of the singular points, is $5U-2s$. We have $5U-2s+U=2(3U-s)$, in accord with the fact that two complementary cubic curves are together the intersection of the cubic surface with a quadric.

Ex. 3. A conic on the cubic surface obtained by a plane drawn through the line a, is associated with the symbol $3U-s-A_1$, and leads, in the plane (ξ, η, ζ), to a cubic curve passing through all the singular points, but having a double point at A_1. A conic, on the cubic surface, in a plane drawn through the line a', is associated with the symbol $U - A_1$, and corresponds to a line, in the plane (ξ, η, ζ), passing through the point A_1.

Ex. 4. On the cubic surface there are seventy-two families of cubic curves, each having as chords the six lines of one row of one of the double-sixes. For the row consisting of the lines $4', 5', 6', c_{23}, c_{31}, c_{12}$ (see above, p. 161), the corresponding family of cubic curves is represented by $2U - A_4 - A_5 - A_6$, giving, in the plane (ξ, η, ζ), conics passing through the three singular points A_4, A_5, A_6. For the row consisting of $1, 2, 3, c_{56}, c_{64}, c_{45}$, the cubic curves are represented by $4U - s - A_1 - A_2 - A_3$. For the row consisting of $2, 2', c_{13}, c_{14}, c_{15}, c_{16}$, the cubic curves are represented by $3U - s + A_1 - A_2$. All other cases are easily deducible from these.

Ex. 5. Two curves of the cubic surface associated, respectively, with the symbols $\lambda U - \Sigma\lambda_r A_r$, $\mu U - \Sigma\mu_r A_r$, meet one another in $\lambda\mu - \Sigma\lambda_r\mu_r$ points. Further, a curve of the preceding symbol is one of a family of which each curve is identified by the assignment of $\frac{1}{2}(n + m)$ points of it, so that the family is said to be of this *freedom*. Also the curves of this family are of *genus* $1 + \frac{1}{2}(n - m)$. Here $n = \lambda^2 - \Sigma\lambda_r^2$, $m = 3\lambda - \Sigma\lambda_r$, and the genus, π, of a curve, without actual multiple points, may be defined by saying that, of the intersections of a surface therewith, if the surface be of sufficiently high order, all but π can be chosen at will. For these matters, and the establishment of the general ideas, reference may again be made to a paper, *Proc. Lond. Math. Soc.* XI, 1912, p. 287; but see, also, Clebsch, *Die Geometrie auf der Flächen dritter Ordnung*, *Crelle's J.*, LXV, 1866.

Ex. 6. In the preceding theory, the cubic curves in the plane (ξ, η, ζ) which represent the plane sections of the cubic surface have been supposed to arise from the cubic surface. We may, however, start with six quite arbitrary points of the plane, and, taking the general cubic curve passing through these, whose equation will be of the form $A\vartheta + B\phi + C\psi + D\chi = 0$, consider the locus of a point in three dimensions whose coordinates are given by

$$x : y : z : t = \vartheta : \phi : \psi : \chi.$$

This locus will be a cubic surface, as the point (ξ, η, ζ), by which $\vartheta, \ldots, \chi$ are expressed, describes the plane. For $\vartheta, \phi, \psi, \chi$ we may then take any four particular cubic polynomials vanishing in the six given points, which are linearly independent of one another, the consequent modification of x, y, z, t being only a linear transformation of these. A particular case arose in Ex. 1 above, where the cubic curves were formed with the help of conics. Another case arises by remarking that there are two cubic curves passing through five of the given points which both have a double point at the remaining point. If we take ξ, η so that these both vanish at this sixth point, but not at any of the other five, the equations of these two cubic curves may be supposed to be $f=0$, $f_1=0$, where $f=a\xi^2+2h\xi\eta+b\eta^2$, and $f_1=a_1\xi^2+2h_1\xi\eta+b_1\eta^2$, in which a, h, b, a_1, h_1, b_1 are all linear homogeneous polynomials in ξ, η, and the third coordinate, ζ. From these we deduce two other cubic polynomials, $(af_1-a_1f)/\eta$ and $(fb_1-f_1b)/\xi$, which, equated to zero, represent cubic curves passing simply through all the six points. The four cubics are linearly independent, since an identity

$$A(af_1-a_1f)/\eta+B(fb_1-f_1b)/\xi-Cf_1+Df=0,$$

in which A, B, C, D are constants, involving the identity

$$f/f_1=(Aa\xi-Bb\eta-C\xi\eta)/(Aa_1\xi-Bb_1\eta-D\xi\eta),$$

would require f, f_1 to have a common factor linear in ξ, η, ζ. We thus obtain the cubic surface expressed by the formulae

$$x=(af_1-a_1f)/\eta,\quad y=(fb_1-f_1b)/\xi,\quad z=-f_1,\quad t=f,$$

the point, (ξ, η, ζ), corresponding to any point, (x, y, z, t), of this, being obtainable by solving the equations

$$az+a_1t+\eta x=0,\quad bz+b_1t-\xi y=0,$$

which are linear in ξ, η, ζ.

Ex. 7. Given seven lines of which no two intersect, having a common transversal, there exists a single point such that the planes joining this to the eight lines touch a quadric cone (p. 150, above, Ex. 4). Given six skew lines with a common transversal, the locus of a point, such that the planes joining it to the seven lines touch a quadric cone, is a cubic curve, having the six lines as chords but not meeting the transversal (above, p. 136). Given five skew lines with a common transversal, the locus of a point, such that the planes joining it to the six lines touch a quadric cone, is a cubic surface on which the six lines lie. The analytical proof of this leads naturally to the formulae of Ex. 6.

Let the common transversal be taken for $z=0$, $t=0$, the planes zxy, txy being any planes drawn through this. A quadric cone which touches the line $z=0$, $t=0$ will meet the planes zxy, txy in two conics touching one another on this line, say at the point $(\eta, -\xi, 0, 0)$. The equation of a plane being $lx+my+nz+pt=0$, the tangential equations of these conics are of the respective forms

$$n(l\eta-m\xi)-(al^2+2hlm+bm^2)=0,$$
$$p(l\eta-m\xi)-(a_1l^2+2h_1lm+b_1m^2)=0;$$

thus, if we denote $al^2+2hlm+bm^2$, and $a_1l^2+2h_1lm+b_1m^2$, respectively, by $\phi(l, m)$ and $\phi_1(l, m)$, and denote $\phi(\xi, \eta)$ and $\phi_1(\xi, \eta)$, respectively, by f and f_1, the coordinates, l, m, n, p, of the general tangent plane of the quadric cone which touches the line $z=0$, $t=0$ at the point $(\eta, -\xi, 0, 0)$, satisfy the relation

$$(l\eta - m\xi)(nf_1 - pf) - [\phi(l, m) f_1 - \phi_1(l, m) f] = 0.$$

This, however, breaks up into the equation $l\eta - m\xi = 0$, and the equation

$$l\frac{af_1 - a_1 f}{\eta} + m\frac{fb_1 - f_1 b}{\xi} - nf_1 + pf = 0,$$

which is thus the equation of the vertex of the cone. Herein the quantities $\xi, \eta, a, h, b, a_1, h_1, b_1$ enter homogeneously to the third order in the aggregate, the expression $(af_1 - a_1 f)/\eta$, for example, being of the first order in ξ and η, and of the second order in the other six parameters.

Now, any line meeting the transversal $z=0, t=0$ is given by two planes $px + qy + rz = 0$ and $px + qy + st = 0$, so that the coordinates of the general plane containing this line are of the forms $(\lambda+\mu)p$, $(\lambda+\mu)q$, λr, μs. For this to be a tangent plane of the quadric cone we require

$$\lambda r(p\eta - q\xi) - (\lambda+\mu)\phi(p, q) = 0, \quad \mu s(p\eta - q\xi) - (\lambda+\mu)\phi_1(p, q) = 0;$$

thus, the condition for the cone to touch the line is

$$-p\eta + q\xi + \frac{1}{r}\phi(p, q) + \frac{1}{s}\phi_1(p, q) = 0.$$

This is a linear condition for the parameters $\xi, \eta, a, h, b, a_1, h_1, b_1$. Thus, when the cone is to touch seven such lines, beside their transversal, it is determinate, in general. When the cone is to touch six such lines, and their transversal, the six parameters, $a, h, \ldots, b_1$ are, in general, definite linear homogeneous functions of ξ, η; and the vertex of the cone, whose coordinates are then homogeneous cubic polynomials in ξ, η, describes a cubic curve. It may be shewn that this line has the six skew lines as chords, but does not meet their transversal; it has already been remarked (Ex. 8, p. 135) that the chords of a cubic curve which meet a line are projected, from a point of the curve, by planes which touch a quadric cone touching the line. Lastly, when the cone is to touch five skew lines and their common transversal, the six parameters $a, h, \ldots, b_1$ are expressible linearly by ξ, η and a third parameter, say ζ; the vertex of the cone then describes a cubic surface, upon which the five lines and their transversal may be shewn to lie.

Derivation of a plane quartic curve from a cubic surface. The converse of the result of the preceding example (7) is that, with the notation we have employed, the six planes joining an arbitrary point of a cubic surface to the lines b, c, d, e, f, of one row of a double-six, and to the line, a', of the other row, which meets all these, touch a quadric cone. This result finds its proper place in the considerations to which we now pass.

First, it must be made clear what is meant by the *tangent plane* of the cubic surface at any point, and by the *inflexional lines* of the surface at this point. If the equation of the surface be $f=0$, the three points in which the line joining two points (x_0, y_0, z_0, t_0), (x, y, z, t), meets the surface, are obtainable by solving the cubic equation in λ which arises by substituting in the equation $f=0$, for the coordinates, respectively $x_0+\lambda x$, $y_0+\lambda y$, etc. If D denote the operator $x\partial/\partial x_0+y\partial/\partial y_0+z\partial/\partial z_0+t\partial/\partial t_0$, and f_0 denote what f becomes when x_0, y_0, z_0, t_0 are put for the coordinates, this equation is $f_0+\lambda Df_0+\frac{1}{2}\lambda^2D^2f_0+\frac{1}{6}\lambda^3D^3f_0=0$. If (x_0, y_0, z_0, t_0) be on the cubic surface, or $f_0=0$, one root of this cubic equation is zero, for all points (x, y, z, t); namely, every line through (x_0, y_0, z_0, t_0) meets the surface once at that point. If (x, y, z, t) lie on the plane whose equation is $Df_0=0$, every line drawn in this plane, through the point (x_0, y_0, z_0, t_0), meets the surface twice at this point. This is the plane which is called the tangent plane of the surface at this point; the curve in which it meets the cubic surface has a double point at this point. Now consider the quadric surface whose equation is $D^2f_0=0$; it is easy to prove that it passes through the point (x_0, y_0, z_0, t_0), and has, for tangent plane at this point, the tangent plane of the cubic surface. This plane therefore contains two lines of this quadric. But when a point, (x, y, z, t), satisfies both $Df_0=0$ and $D^2f_0=0$, the above cubic equation in λ has three zero roots. We see thence that the generators at (x_0, y_0, z_0, t_0), of the quadric $D^2f_0=0$, meet the cubic surface in three coincident points at (x_0, y_0, z_0, t_0); these are then the tangent lines at this point of the two branches of the section of the cubic surface by its tangent plane. They are the lines called the inflexional lines of the surface at (x_0, y_0, z_0, t_0); they lie in the tangent plane at this point. More generally, when (x_0, y_0, z_0, t_0) is not necessarily a point of the cubic surface, the plane $Df_0=0$ is called the polar plane of (x_0, y_0, z_0, t_0) in regard to the cubic surface, and the quadric $D^2f_0=0$ is called the polar quadric; it is easy to see that, if Δ denote what the operator D becomes when x, y, z, t are interchanged, respectively, with x_0, y_0, z_0, t_0, the polar plane can be given by $\Delta^2f=0$, and the polar quadric by $\Delta f=0$; thus, if a point lie on the polar plane of another point, this lies on the polar quadric of the first. In particular, the polar quadric of a point (x_0, y_0, z_0, t_0) meets the cubic surface in points at which the tangent plane passes through (x_0, y_0, z_0, t_0). Further, it may be remarked that if two curves on a cubic surface intersect in a point O, the plane determined by the tangent lines of these curves at O is the tangent plane of the surface.

We take now a point, O, of the cubic surface, not lying on any of the lines of the surface. Then we consider the curve of the surface which is the locus of points, P, of the surface, whereat the tangent plane passes through O; as we have seen, this is the intersection of the cubic surface with the polar quadric of O; moreover, this polar quadric passes through O, has the same tangent plane as the cubic surface at O, and contains the two inflexional lines of the surface at O. The curve in question, being the intersection of the cubic surface with a quadric, is of the sixth order; by what is said, it passes through O, having there a double point, its two tangent lines at this point being the inflexional lines of the cubic surface. Thus an arbitrary plane through O meets the curve in four other points, and the curve is projected from O, on to an arbitrary plane, into a quartic curve. Conversely, it may be stated, any general quartic curve in a plane can be shewn to be capable of being thus derived. We shall denote the sextic curve on the cubic surface by ϑ.

If an arbitrary quadric cone be drawn with vertex O, four of its twelve points of intersection with ϑ will be at O, and there will be eight others; it may happen that these eight other points coincide in pairs, the quadric cone touching the curve ϑ in four points, not at O. When this is so, we may, for brevity, speak of the cone as a contact cone. In such case, if a further arbitrary quadric cone be drawn, with vertex at O, to pass through the four points of contact, with the curve ϑ, of the contact cone in question, the remaining four intersections, with ϑ, of this further cone, are also points of contact with ϑ, of a contact cone; this we speak of as being of the same *system* as the former. There are in all sixty-three different systems of contact cones, with the same vertex, O. In particular such a contact cone may consist of two planes; then either it consists of the tangent plane of the cubic surface at O, taken with the plane joining O to one of the twenty-seven lines of the surface; or it consists of the planes joining O to two non-intersecting lines of the surface. There are 216 pairs of non-intersecting lines of the surface, but it will be seen that these arrange themselves in sets of six pairs giving degenerate contact cones of the same system.

We approach the proof of these statements by considering, first, degenerate contact cones consisting of the tangent plane at O taken with the plane joining O to a line of the surface. When the curve ϑ is projected from O on to a plane, ϖ, the tangent line of the resulting quartic curve, at any point, is evidently the intersection with ϖ of the tangent plane of the cubic surface at the corresponding point, P, of the curve ϑ, this tangent plane, by hypothesis, passing through O. In particular, by taking P at O, the tangent plane of the cubic surface at O gives rise to a tangent line of the quartic curve; but there are two ways of approach of P to O, along the two branches at O of the curve ϑ; thus the tangent plane of the cubic surface at O touches the plane quartic curve in two points, lying, namely, on the inflexional lines of the cubic surface at O. The plane joining O to any one, say l, of the lines of the cubic surface, equally gives a double-tangent of the plane quartic curve; the points of contact are the projections of the two points of the line l at which the plane Ol is a tangent plane of the cubic surface, namely the points in which l is met by the conic in which the plane Ol cuts the surface. We thus see that the plane quartic curve has twenty-eight double-tangents. Now, let any plane be drawn through the line l, meeting the cubic surface also in a conic, say σ. The six intersections of this plane with the curve ϑ will contain the two points of this we have just found lying on the line l, which do not lie on σ, in general; and there will be four other points, which will lie on the conic σ. We have remarked that, when two curves of the cubic surface have distinct tangent lines at a point where the curves intersect, the plane of these lines is the tangent plane of the surface. Thus, the tangent plane of the cubic surface at any one of the four intersections of the curve ϑ and the conic σ, touches both these curves; while, by definition of ϑ, this plane passes through the point O. The quadric cone joining O to the conic σ is thus a contact cone, touching the curve ϑ in four points. This cone will meet the plane ϖ in a conic having four points of contact with the plane quartic curve, and, therefore, no further intersections with this. We have remarked that, in general, the conic σ does not contain the two points where the curve ϑ meets the line l; the curve ϑ, however, lies on a quadric, the polar quadric of O; the section of this quadric by the plane of σ is a conic; this conic contains the two points where ϑ meets l, and the four points where ϑ meets σ. This conic also meets the two generators at O of the polar quadric of O. Thus, on projection to the plane ϖ, the four points of contact of the conic arising by projection of σ, with the plane quartic curve, lie on a conic whose other four intersections with the quartic are the points of contact with this of two of its double-tangents. Of the contact cones, $O\sigma$, joining O to a conic σ, there is an infinite number, each defined by a plane through the line l; and, as an arbitrary line through O meets the cubic surface again in two points, two such cones pass through this line. In particular, there are five planes through the line l for which the conic σ breaks up into two lines, and the cone $O\sigma$ becomes two planes; we may also consider the plane-pair consisting of the plane Ol and the tangent plane at O as a sixth degenerate contact cone. Thus, in the plane ϖ, for the plane quartic curve, the result is that two of the contact conics, of the system derived from the line l, pass through an arbitrary point of the plane, and six of them break up into two lines, each a bitangent of the plane quartic curve. This system of contact conics may remind us of the conics in a plane which touch four given lines, which is indeed a particular case, arising when the quartic curve breaks up into four lines. In the particular case, all the conics are inpolar to any conic passing through three particular points, which form a triad self-polar in regard to all the conics touching the four lines. In the general case now before us, it is also true that the contact conics of the system considered are all inpolar to a conic, which is unique; in special this is true of the six contact conics which break up into two lines. Thus, as a pair of lines inpolar to a conic intersect thereon, we have the result that the six points of intersection of the pairs of bitangents which are particular contact conics of the system, lie upon a conic, to which all the proper contact conics of the system are inpolar. Such a set of six pairs of bitangents is called a Steiner system; there are evidently twenty-seven systems arising in this way. The proof of several of the statements here made has not been given; but the character of this proof will be clear after the consideration of the thirty-six cases of contact conics associated with pairs of non-intersecting lines of the cubic surface, to which we now pass.

Consider one of the thirty-six double-sixes of lines of the cubic surface; with this, as we have seen, are associated two families of cubic curves on the surface. Through the point O, and an arbitrary point, say H, of the curve ϑ, a definite curve of one of these families can be drawn; this curve is projected from O by a quadric cone, whose residual intersection with the cubic surface is, as we have seen, a cubic curve of the other family; and these two curves cross one another at five points, as was also proved. Of these five points, O is evidently one; at any other point of crossing of these curves, which lie both on the cone and on the cubic surface, the plane of the tangent lines of the two curves is as well a tangent plane of the cone, passing through O, and a tangent plane of the cubic surface; thus the point of crossing lies on the curve ϑ. Wherefore, the quadric cone touches the cubic surface, and the curve ϑ, at the assumed point H, and also at three other points, say P, Q and R, and is such a contact cone as was in view. It may be remarked, too, that, in virtue of what has previously been proved (above, p. 177), the polar line of OH, in regard to the Schur quadric belonging to the double-six under consideration, lies in the plane PQR, the five points O, H, P, Q, R forming a self-conjugate pentad in regard to this quadric, while the contact cone is outpolar to this quadric. The system of contact cones, of vertex O, of the curve ϑ, obtained in this way, consists of the cones projecting from O the cubic curves, through O, of either of the two families belonging to the double-six; and, since an arbitrary line through O meets the cubic surface in two further points, and a cubic curve of the family is determined by O and another point of the surface, we see that there are two such cones containing, as generator, an arbitrary line drawn through O. Now, suppose that a and a' are two lines of different rows of the double-six, not intersecting one another; then, one cubic curve, of those associated with the double-six, consists of the line a', together with the conic in which an arbitrary plane drawn through the other line, a, meets the cubic surface, other than a itself; a degenerate cubic of the complementary family is, similarly, constituted by the line a taken with a conic of the cubic surface lying in a plane through a'. Hence, a degenerate contact cone is constituted by the two planes Oa, Oa'. Also, if the conics in the planes Oa, Oa' meet a and a', respectively, in A, B and A', B', it is easily seen, in accordance with a remark made above, that A, B, A', B' form a self-polar tetrad in regard to the Schur quadric belonging to the double-six. On the other hand, the contact cones from O are all inpolar to a definite quadric cone with O as vertex. For, such a contact cone is determined by one of the cubic curves, associated with the double-six, which pass through O; and this curve is determined by its tangent line at O. The equation of the cone thus contains rationally a parameter, determining this tangent line in the tangent plane at O of the cubic surface. We have remarked, however, that there are two of these contact cones possible, containing an arbitrary line drawn through O. The parameter thus enters to the second order in the equation of the tangent cone; this equation is, therefore, of the form $\lambda^2 U + \lambda V + W = 0$, where λ is the parameter, and U, V, W are cones with vertex at O. By considering the tangential equation of this cone, we see that there is a single definite cone, with O as vertex, which is outpolar to all the contact cones. In particular, this single cone, say Ω, is outpolar to all the plane-pairs such as Oa, Oa'. Hence we have the result that if a, a'; b, b'; ..., f, f' be the pairs of lines in the six *columns* of a double-six, and O be an arbitrary point of the cubic surface, the six transversals from O, to a and a'; to b and b'; etc., lie upon a quadric cone, Ω.

Ex. 1. A similar argument is applicable to the case of the contact cones obtained by joining O to the conics of the cubic surface lying in planes through a line, l, of the surface. Hence we have the result that, if $C_1, C_2, \ldots, C_5$ be the points of intersection of the five pairs of lines of the cubic surface, of which a pair lie in a plane through the line l, then the quadric cone containing the five lines OC_r contains also the line in which the plane Ol meets the tangent plane at O. It will be recalled that it has been proved that the five planes joining O to one from each of the five pairs of intersecting lines meeting l, touch a quadric cone which touches the plane Ol (above, p. 195).

It is known, too, that the lines joining the pairs of the five points C_r, and the planes containing the threes of these, meet an arbitrary plane in ten points, and ten lines, of which each point is the pole of one of the lines in regard to a certain conic in that plane (Vol. II, p. 218). When the plane is any plane through the line l, the conic in question is the conic of the cubic surface which lies in this plane; and any cubic curve through the five points C_r meets this plane in points which form a self-conjugate triad in regard to this conic (A. C. Dixon, *Quart. J. of Math.*, XXIII, 1889, p. 343).

Ex. 2. We may deal with the theory of the contact cones of the curve ϑ with analysis. It will be sufficient here to consider the case of contact cones associated with a double-six. As before, let O be any point of the cubic surface, and a, a' be two non-intersecting lines of the surface. Let the conics of the surface which lie in the planes Oa, Oa' meet the lines a, a', respectively, in A, B and A', B'. Choose coordinates so that the planes Oa, Oa', OAA' are, respectively, $y=0$, $z=0$, $x=0$, the plane $t=0$ being any plane through the line AA'. The line a, in the plane $y=0$, will have an equation $x+\lambda t=0$; as it does not intersect the line a', it must intersect the conic of the cubic surface which lies in the plane Oa', or $z=0$, the point of intersection being on the line $y=0$, $z=0$; this conic may then be taken to have the equations

$$z=0, \quad x^2+\lambda xt+mxy+nyt=0.$$

The line a' being, similarly, $z=0$, $x+\mu t=0$, the conic in the plane Oa has equations of the forms $y=0$, $x^2+\mu xt+rxz+kzt=0$. Hence we find, for the equation of the plane OBB', $\xi=0$, where $\xi=(\lambda-\mu)x+y(n-m\mu)+z(r\lambda-k)$; and, for the equation of the tangent plane of the cubic surface at O, $P=0$, where

$$P=\lambda\mu x+n\mu y+k\lambda z.$$

Then the equation of the cubic surface is $C=0$, where

$$C=t^2P+tQ+R,$$

in which

$$Q=(\lambda+\mu)x^2+(n+m\mu)xy+(k+r\lambda)xz+2fyz,$$
$$R=x^2(x+my+rz)+yz(ux+vy+wz),$$

f, u, v, w being constants. Taking, then, $U=2tP+Q$, so that $U=0$ is the polar quadric of O, we can verify the identity

$$4PC=U^2-x^2\xi^2+4yz\psi,$$

where

$$\psi=x^2(n-m\lambda)(r\mu-k)+P(ux+vy+wz)-fQ+f^2yz.$$

This identity is the same as

$$4PC = U^2 - \Phi^2 + 4yz\Psi,$$

wherein $\Phi = x\xi + 2\lambda yz$, $\Psi = \psi + \lambda x\xi + \lambda^2 yz$, and λ is an arbitrary parameter. The curve ϑ, for which $C = 0$, $U = 0$, thus lies on the quartic cone, of vertex O, given by

$$\Phi^2 - 4yz\Psi = 0, \text{ or } x^2\xi^2 - 4yz\psi = 0,$$

and $\Phi = 0$ represents the general quadric cone, of vertex O, containing the points A, B, A', B'. A general contact cone is therefore given by $\Psi = 0$.

It is clear from the first form of the identity that the points of the curve given by the intersection of the quadrics

$$U + x\xi + 2\lambda yz = 0,\ U - x\xi - 2\lambda^{-1}\psi = 0$$

lie either on the cubic surface, or on the tangent plane, $P = 0$. It can be seen that these quadrics have in common a line, through O, lying in the tangent plane, variable with λ. The common curve of these quadrics, however, lies on the cone $\Psi = 0$, which thus contains a cubic curve, of the cubic surface, passing through O. Another such cubic curve is obtained from the quadrics

$$U - x\xi - 2\lambda yz = 0,\ U + x\xi + 2\lambda^{-1}\psi = 0.$$

Further elementary results will be found in papers, *Proc. Lond. Math. Soc.*, IX, 1910, p. 145, and p. 205, the latter of which gives valuable references to the literature. The theory of the contact cones was originally obtained with the help of multiple theta functions, and belongs to any curve whatever which is not rational or elliptic. See Clebsch, *Crelle's J.*, LXIII, 1864, p. 189; and, for a general introduction, the writer's *Abel's Theorem* (Cambridge, 1897), pp. 377—392. The result we have proved that, of the points of contact, H, P, Q, R, of a contact cone with the curve ϑ, the three latter are determined when the first is given arbitrarily, arises from the fact that the curve ϑ is of *genus* 3 (see above, p. 194). The number 4, of points of contact, is the value of $2p-2$ when $p=3$; the number 63, of systems of contact cones, is the value of $2^{2p}-1$ when $p=3$; the number 6, of degenerate contact cones of any system is the value of $2^{p-2}(2^{p-1}-1)$ when $p=3$. For a general sextic curve obtained by the intersection of any cubic surface with any quadric surface, there are, similarly, quadrics which touch the curve in six points, the value of the genus, p, in this case, being 4. An interesting geometrical introduction to this case is given by Milne, *Proc. Lond. Math. Soc.*, XXI, 1922, p. 373; see also P. Roth, *Monatsh. f. Math. u. Phys.*, XXII, 1911, pp. 64—88.

The equation of the cubic surface referred to two Steiner trihedrals. Consider a cubic surface defined by the vanishing of a general cubic polynomial in the coordinates x, y, z, t, and assume that this surface contains at least one line, and thence that there are five planes through this line whose further intersection with the surface consists of a pair of lines (as above, p. 167); let the coordinates be taken, with the point $x = 0$, $y = 0$, $z = 0$ on the surface, so that one line of the surface is $x = 0$, $t = 0$, so that another pair of lines of the surface which meet this are $t=0$, $y=0$ and $t=0$, $z=0$, and so that a further plane, containing a pair of lines beside $x=0$, $t=0$, is given by $t+x=0$, these lines being, respectively, $t+x=0$, $v=0$, and $t+x=0$, $w=0$, where v, w are homogeneous polynomials in x, y, z. Then the equation of the cubic surface is necessarily of the form $t^2P+t\phi+xyz=0$, where ϕ is a homogeneous quadratic polynomial in x, y, z, and P is linear in these; and, as $t+x=0$ intersects the surface in the lines for which $v=0$, $w=0$, we have $\phi=xP+yz-vw$. The equation is, therefore, $t^2P+t(xP+yz-vw)+xyz=0$. It is at once seen, however, that this is identically the same as

$$(t+x)(t+by)(t+cz)=t(t+x+qv)(t+x+rw),$$

in which b, c, q, r are constants, provided that these constants can be so chosen that $bc=qr$, and also so that there is the identity

$$x+qv+rw-by-cz+bcP=0;$$

if we assume general values for v, w, P, as linear functions of x, y, z, write $b=\lambda q$, $c=\lambda^{-1}r$, and equate coefficients of x, y, z in this identity, two of the resulting equations determine q and r uniquely in terms of λ, and the substitution of these in the third equation leads to an equation for λ of the fourth order. Thus the form in question is possible. In more general terms, the equation of the cubic surface is reducible to the form $UVW=U'V'W'$, in which $U=0$, ... are six planes. By considering the geometrical meaning of this (as will appear more particularly in a moment), we see that these are the planes of a Steiner trihedral pair, previously explained (above, pp. 164, 174); so that there are 120 ways in which the equation of the cubic surface can be reduced to this form.

Now assume this form; and let $U_0, V_0, \ldots$ be the values of $U, V, \ldots$, respectively, at an arbitrary general point, (x_0, y_0, z_0, t_0), of the surface, which does not lie on a line of the surface; put $u=U/U_0$, $v=V/V_0$, etc.; then, in virtue of $U_0V_0W_0=U_0'V_0'W_0'$, we can suppose the equation of the surface to be $uvw=u'v'w'$. The six planes $u=0$, $v=0$, ... must be expressible by four of them; thus we can suppose the existence of two identities of the forms

$$a_1u+b_1v+c_1w=a_1'u'+b_1'v'+c_1'w',$$
$$a_2u+b_2v+c_2w=a_2'u'+b_2'v'+c_2'w',$$

wherein, since $u, v, \ldots$ reduce to unity at (x_0, y_0, z_0, t_0), the constants are such that

$$a_1+b_1+c_1=a_1'+b_1'+c_1', \quad a_2+b_2+c_2=a_2'+b_2'+c_2',$$

and, therefore, such that

$$(a_1+\sigma a_2)+(b_1+\sigma b_2)+(c_1+\sigma c_2)=(a_1'+\sigma a_2')+(b_1'+\sigma b_2')+(c_1'+\sigma c_2'),$$

whatever σ may be. We may, however, in three ways, choose σ so that

$$(a_1 + \sigma a_2)(b_1 + \sigma b_2)(c_1 + \sigma c_2) = (a_1' + \sigma a_2')(b_1' + \sigma b_2')(c_1' + \sigma c_2').$$

But, when we have six quantities $\lambda, \mu, \nu, \lambda', \mu', \nu'$ which satisfy the two relations $\lambda + \mu + \nu = \lambda' + \mu' + \nu'$, and $\lambda\mu\nu = \lambda'\mu'\nu'$, if we put

$$\alpha = (\mu' - \nu)/(\lambda - \lambda'), \quad \beta = (\nu' - \lambda)/(\mu - \mu'), \quad \gamma = (\lambda' - \mu)/(\nu - \nu'),$$

and, then, $\alpha' = 1 - \alpha$, $\beta' = 1 - \beta$, $\gamma' = 1 - \gamma$, it can at once be verified that $\lambda, \mu, \nu, \lambda', \mu', \nu'$ are, respectively, proportional to $\beta\gamma', \gamma\alpha', \alpha\beta', \beta'\gamma, \gamma'\alpha, \alpha'\beta$. Hence, in our case, the identities connecting $u, v, \ldots$ are capable of the forms

$$\beta_r\gamma_r'u + \gamma_r\alpha_r'v + \alpha_r\beta_r'w = \beta_r'\gamma_r u' + \gamma_r'\alpha_r v' + \alpha_r'\beta_r w',$$

for $r = 1, 2, 3$, these being equivalent to two independent equations. If, then, we put

$$\xi_r = \beta_r\gamma_r'u, \quad \eta_r = \gamma_r\alpha_r'v, \quad \zeta_r = \alpha_r\beta_r'w,$$
$$\xi_r' = \beta_r'\gamma_r u', \quad \eta_r' = \gamma_r'\alpha_r v', \quad \zeta_r' = \alpha_r'\beta_r w',$$

the equation of the cubic surface is capable of the forms

$$\xi_r\eta_r\zeta_r = \xi_r'\eta_r'\zeta_r',$$

where $\xi_r, \eta_r, \ldots$ are subject to two linear identities, of which one is

$$\xi_r + \eta_r + \zeta_r = \xi_r' + \eta_r' + \zeta_r'.$$

With these equations we can at once specify the equations of the twenty-seven lines of the cubic surface. In the first place, the line of intersection of any one of the planes $\xi_r = 0$, $\eta_r = 0$, $\zeta_r = 0$ with any one of the planes $\xi_r' = 0$, $\eta_r' = 0$, $\zeta_r' = 0$ is evidently a line of the surface, the same line for the three values of the suffix r, since the plane $\xi_r = 0$ is the same as $u = 0$, etc. Denoting such a line by $\xi\xi'$, $\xi\eta'$, etc., we may represent these lines, respectively, with a notation previously employed, by the scheme

$$\begin{array}{lll} \xi\xi', & \xi\eta', & \xi\zeta' \\ \eta\xi', & \eta\eta', & \eta\zeta' \\ \zeta\xi', & \zeta\eta', & \zeta\zeta' \end{array} \equiv \begin{array}{lll} c_{12}, & 1, & 2' \\ 1', & c_{31}, & 3 \\ 2, & 3', & c_{23} \end{array}$$

and may thence recover the original definition of the trihedral pair. In the next place, if X_r', Y_r', Z_r', respectively, denote $\xi_r', \eta_r', \zeta_r'$, in some order, the line given by $\xi_r = X_r'$ and $\eta_r = Y_r'$ is, by the identity connecting the six planes, also on the plane $\zeta_r = Z_r'$; and thus lies on the surface. This line may be represented by $(X', Y', Z')_r$; by taking all three values of r, and all permutations for X_r', Y_r', Z_r', so that, for example, $(\xi', \zeta', \eta')_r$ stands for $\xi_r = \xi_r'$, $\eta_r = \zeta_r'$, $\zeta_r = \eta_r'$, we obtain eighteen lines. With the previous notation they may be identified as follows:

$(\eta', \xi', \zeta')_r \equiv c_{14}, c_{15}, c_{16}$; $(\zeta', \eta', \xi')_r \equiv c_{24}, c_{25}, c_{26}$; $(\xi', \zeta', \eta')_r \equiv c_{34}, c_{35}, c_{36}$
$(\xi', \eta', \zeta')_r \equiv c_{56}, c_{64}, c_{45}$; $(\zeta', \xi', \eta')_r \equiv 4, 5, 6$; $(\eta', \zeta', \xi')_r \equiv 4', 5', 6'$,
where the meaning is, that, in each case, the three lines on the right arise, respectively, for $r = 1, 2, 3$.

The Cremona form of the equation of the cubic surface. Remarking the identity

$$(x+y+z)^3 = x^3 + y^3 + z^3 + 3(y+z)(z+x)(x+y),$$

and putting

$$\xi = x_2 + x_3, \quad \eta = x_3 + x_1, \quad \zeta = x_1 + x_2,$$
$$\xi' = -(x_5 + x_6), \quad \eta' = -(x_6 + x_4), \quad \zeta' = -(x_4 + x_5),$$

or

$$2x_1 = \eta + \zeta - \xi, \; \ldots, \; 2x_4 = \xi' - \eta' - \zeta', \; \ldots,$$

the two equations $\xi\eta\zeta = \xi'\eta'\zeta'$, $\xi + \eta + \zeta = \xi' + \eta' + \zeta'$, lead to the two

$$x_1^3 + x_2^3 + x_3^3 + x_4^3 + x_5^3 + x_6^3 = 0,$$
$$x_1 + x_2 + x_3 + x_4 + x_5 + x_6 = 0.$$

The cubic surface may then be supposed to be given by these two equations, taken with another necessary linear equation connecting $x_1, \ldots, x_6$. (Cremona, *Math. Annal.*, XIII, 1878.)

The planes $x_1 = 0, \ldots, x_6 = 0$, save for permutations, are determinate, in association with a particular double-six of lines of the cubic surface. For fifteen of the lines of the surface are given each by a set of three planes such as $x_1 + x_2 = 0$, $x_3 + x_4 = 0$, $x_5 + x_6 = 0$. The other twelve lines, which are $5, 5', c_{6m}$ and $6, 6', c_{5m}$ (for $m = 1, 2, 3, 4$), form a double-six, and can be found in terms of the coefficients in the remaining equation connecting $x_1, \ldots, x_6$, by the solution of a quadratic equation.

The Sylvester form of the equation of the cubic surface. The Cremona form may be regarded as a particular case of a form $\Sigma a_r x_r^3 = 0$, subject to the relation $\Sigma x_r = 0$, with another linear relation connecting $x_1, \ldots, x_6$; that namely in which the coefficients $a_1, \ldots, a_6$ are equal. As Reye has shewn (*Crelle's J.*, LXXVIII, 1874), there exists a set of six planes $x_1 = 0, \ldots$, corresponding to any arbitrary line, for which the more general form is possible. We assume this form, in order to reach a remarkable form, as a sum of *five* cubes, due to Sylvester (1851; *Math. Papers*, I, p. 195); the method we follow can in particular be carried out starting from the Cremona form. It will be seen that it is supposed that the cubic surface is quite general.

We assume, that is, the cubic surface given by the equations

$$\sum_{r=1}^{6} a_r x_r^3 = 0, \quad \sum_{r=1}^{6} x_r = 0, \quad \sum_{r=1}^{6} h_r x_r = 0,$$

of which the last is the second necessary condition connecting $x_1, x_2, \ldots$; in this we may, evidently, replace $h_1, h_2, \ldots$ by $h_1+\lambda$, $h_2+\lambda, \ldots$, where λ is arbitrary. We take the most general case possible, assuming that no two of $h_1, h_2, \ldots, h_6$ are equal; this is necessary for the method we employ. Then, $\phi(\sigma)$ denoting the product of the six factors $\sigma - h_r$, and $\phi'(\sigma) = d\phi(\sigma)/d\sigma$, none of the quantities $\phi'(h_r)$ is zero. We can also suppose that the sum, of six terms, $\Sigma a_r^{-1}\phi'(h_r)$, does not vanish. For, the quintic polynomial in λ expressed by $\Sigma a_r^{-1}\phi'(h_r+\lambda)$ vanishes only for five values of λ, unless it vanish identically—and that is not so; this we see by equating to zero the coefficients of the various powers of λ, and eliminating, from the six resulting equations, the six coefficients a_r^{-1}; the result is the product of the differences of $h_1, \ldots, h_6$. To ensure that $\Sigma a_r^{-1}\phi'(h_r)$ does not vanish we have, then, only to increase each of $h_1, h_2, \ldots$ by a suitable value of λ; this we suppose to be done. Now we consider the expression

$$\psi(\sigma), = -\phi(\sigma) \sum_{r=1}^{6} \frac{a_r^{-1}\phi'(h_r)}{\sigma - h_r};$$

by the condition just secured, this is a quintic polynomial in σ; let its roots be $k_1, \ldots, k_5$. It is necessary for the method we employ to assume that *no two of these roots are equal, nor any of them equal to any of* $h_1, \ldots, h_6$.

Next, let $$F(\sigma) = \phi(\sigma) \sum_{r=1}^{6} \frac{x_r}{\sigma - h_r},$$

so that, by the conditions connecting $x_1, \ldots, x_6$, this is a cubic polynomial in σ. The equation $F(\sigma) = 0$, when $x_1, \ldots, x_6$, as well as σ, are variable, is the equation of a plane of a cubic developable; as $F(h_r) = \phi'(h_r)\, x_r$, this is the cubic developable of which $x_1 = 0$, $x_2 = 0, \ldots, x_6 = 0$ are planes.

Now consider, as depending upon σ, the fraction

$$[F(\sigma)]^3/\phi(\sigma)\,\psi(\sigma),$$

of which the numerator is of the ninth order, and the denominator is of the eleventh order, and has its roots all different. Expressing this in partial fractions we can infer that

$$\sum_{r=1}^{6} \frac{F^3(h_r)}{\phi'(h_r)\,\psi(h_r)} + \sum_{s=1}^{5} \frac{F^3(k_s)}{\phi(k_s)\,\psi'(k_s)} = 0;$$

as $F(h_r) = \phi'(h_r)\, x_r$ and $\psi(h_r) = -a_r^{-1}[\phi'(h_r)]^2$, the first sum is $-\Sigma a_r x_r^3$. If we put

$$\xi_s = [\phi(k_s)]^{\frac{1}{2}} \sum_{r=1}^{6} \frac{x_r}{k_s - h_r},$$

so that $\xi_s = 0$ represents one of five particular planes of the cubic developable containing the six planes $x_r = 0$, and also put

$$\mu_s = [\phi(k_s)]^{\frac{1}{2}}/\psi'(k_s),$$

we are finally able to infer that

$$\sum_{r=1}^{6} a_r x_r^3 = \sum_{s=1}^{5} \mu_s \xi_s^3,$$

where also $\sum_{s=1}^{5} \mu_s \xi_s = 0$; for

$$\sum_{s=1}^{5} \mu_s \xi_s = \sum_s \frac{\phi(k_s)}{\psi'(k_s)} \sum_r \frac{x_r}{k_s - h_r}, = \sum x_r \sum_s \frac{\phi(k_s)}{\psi'(k_s)(k_s - h_r)},$$

and we have only to consider the expression of $F(\sigma)/\psi(\sigma)$ in partial fractions, regarded as depending on σ. The relation $\Sigma \mu_s \xi_s = 0$ is the only necessary relation among the five planes $\xi_s = 0$.

This method of deducing Sylvester's form was given by Beltrami (*Rendic. Lomb.*, Milan, XII, 1879, p. 24). Sylvester's own enunciation was accompanied by indications of a proof, depending on the properties of the Hessian surface, which has been developed by Clebsch (*Crelle's J.*, LIX, 1861, p. 193); the nature of this proof will be understood from the properties of the Hessian developed below on the basis of Sylvester's form. The reader may also consult Gordan, *Math. Annal.*, V, 1872, p. 341. Also Segre and Sturm, *Archiv d. Math. u. Phys.*, X, 1906, pp. 209, 216. And, generally, Steiner, *Gesamm. Werke*, II, p. 657; Sturm, *Synthetische Untersuch. ü. Fl. dr. Ordn.*, Leipzig, 1867; Cremona, *Mémoire de géom. pure sur les surf. d. trois. ordre*, *Crelle's J.*, LXVIII, 1868, and Grassmann, *Crelle's J.*, XLIX, 1855, p. 37.

The polars in regard to a cubic surface. The Hessian surface. We have already explained (above, p. 182) what is meant by the polar quadric, and by the polar plane, of a point in regard to a cubic surface. In regard to the polar quadrics, it is convenient to notice here, (a), that the polar quadrics of all points of a line are a system of quadrics all passing through the curve of intersection of any two of them, as is clear because the polar quadric of a point is expressed by an equation which is linear in the coordinates of this point; and, (b), that the polar quadrics of all points of a plane have common the eight points of intersection of any three of these quadrics. When the polar quadric of one point passes through another point the polar plane of this other passes through the first point; there are thus eight points of which a given plane is the polar plane. In regard to polar planes, we consider similarly, (a), the envelope of the polar planes of the points of a given line. These planes are expressed by an equation of the form

$$[(\lambda x_1 + \mu x_2)\partial/\partial x + \dots]^2 f = 0,$$

that is, of the form $\lambda^2 P + 2\lambda\mu Q + \mu^2 R = 0$, where P, Q, R are linear in the current coordinates x, y, z, t. These are the tangent planes of a quadric cone, whose equation is $RP = Q^2$. But it may happen, as will be seen, that the cone degenerates, all these polar planes having a line in common. Further, (*b*), we consider the envelope of the polar planes of the points of a given plane, say ϖ. It is clear that the *coordinates* of such a polar plane are homogeneous quadratic functions of *three* variable parameters. From this it follows (as will be seen below, p. 222) that these polar planes all touch a particular surface of the third order, which has four double points. The polar quadrics of the points of this cubic surface, taken in regard to the original cubic surface, can then be proved to touch the plane, ϖ, from which we began.

The polar quadric of a point, (x_0, y_0, z_0, t_0), may be a cone, of vertex (x_1, y_1, z_1, t_1). The condition for this is the coexistence of four equations, linear both in regard to (x_0, y_0, z_0, t_0), and in regard to (x_1, y_1, z_1, t_1), as we easily see; and these equations are symmetrical in regard to both these points. If we denote the values, at (x_1, y_1, z_1, t_1), of $\partial^2 f/\partial x^2, \partial^2 f/\partial x \partial y, \ldots, \partial^2 f/\partial t^2$, wherein $f = 0$ is the equation of the cubic surface, respectively by $f_{11}, f_{12}, \ldots, f_{44}$, these four equations are $x_0 f_{r1} + y_0 f_{r2} + z_0 f_{r3} + t_0 f_{r4} = 0$, for $r = 1, \ldots, 4$. We infer, then, that the necessary and sufficient condition is that both (x_0, y_0, z_0, t) and (x_1, y_1, z_1, t_1) should be points of a quartic surface whose equation is expressible by the vanishing of the determinant, of four rows and columns, whose general element is f_{rs}. This is the Hessian surface of the cubic surface, of which (x_0, y_0, z_0, t_0) and (x_1, y_1, z_1, t_1) are then said to be *corresponding* points. (Cf. p. 188, above.) It is important to recognise that this surface is independent of the choice of coordinates by which it is expressed.

The Hessian may be formed by the rule for any cubic surface, however special. When the surface is sufficiently general to be expressible in Sylvester's form

$$ax^3 + by^3 + cz^3 + dt^3 + eu^3 = 0,$$

with the condition

$$ax + by + cz + dt + eu = 0,$$

the polar plane of a point (X, Y, Z, T, U), where also

$$aX + \ldots + eU = 0,$$

is easily found to be

$$aX^2 x + bY^2 y + cZ^2 z + dT^2 t + eU^2 u = 0;$$

and this equation is also that of the polar quadric of (x, y, z, t, u) when $X, Y, \ldots$ are regarded as current coordinates. Hence the equation of the Hessian surface is

$$ax^{-1} + by^{-1} + cz^{-1} + dt^{-1} + eu^{-1} = 0,$$

the relation connecting corresponding points being

$$x_0 x_1 = y_0 y_1 = z_0 z_1 = t_0 t_1 = u_0 u_1.$$

If $F=0$ denote the Hessian, the tangent plane, at any point $(x_0, y_0, \ldots)$ of this Hessian, calculated by the formula

$$[x\partial/\partial x_0 + y\partial/\partial y_0 + z\partial/\partial z_0 + t\partial/\partial t_0]F = 0,$$

where, in F, the coordinate u_0 has been replaced by

$$-e^{-1}(ax_0 + by_0 + cz_0 + dt_0),$$

may easily be seen to be

$$axx_0^{-2} + \ldots + dtt_0^{-2} + euu_0^{-2} = 0.$$

If $(x_1, y_1, \ldots)$ be the point corresponding to $(x_0, y_0, \ldots)$, this equation is the same as $axx_1^2 + \ldots + dtt_1^2 + euu_1^2 = 0$. Thus the tangent plane of the Hessian surface at any point is the polar plane, in regard to the cubic surface, of the corresponding point of the Hessian. This result may be obtained without the special form.

Now consider the relations of the five Sylvester planes, $x=0, \ldots, u=0$, to the Hessian surface. If these are clear, since the Hessian may be formed from the original form of the cubic surface, the method of reduction of this to the Sylvester form may be inferred. The five planes, $x=0, \ldots, u=0$, meet in threes in ten points. It is easily seen that every line drawn through one of these meets the Hessian in two coincident points there. The Hessian has, therefore, these ten points as double points. The five planes also meet in pairs in ten lines, each of which, evidently, contains three of these double points; these ten lines lie wholly on the Hessian. In any one of the ten planes there are four of the lines, and six of the double points, lying at the intersections of the lines. The polar quadric, in regard to the cubic surface, of any one of the ten points, breaks up into two planes, containing the line of intersection of two of the five planes and harmonic therewith. For instance, the polar quadric of the point $x=0$, $y=0$, $z=0$, is $dT^2t + eU^2u = 0$. By this condition the ten points are definable directly from the equation of the cubic surface.

Ex. 1. Prove that the vertices of a pair of Steiner trihedrals (above, p. 204) are corresponding points of the Hessian surface.

Ex. 2. It has been seen that there are eight points of which a given general plane is the polar plane, in regard to the cubic surface. Prove that, upon the line joining any two of these eight points, there is an infinite number of pairs of points, forming an involution, such that the polar planes, in regard to the cubic surface, of the two points of a pair, are the same. Also, that the polar planes of all points of such a joining line pass through another line; the axial pencil of these planes, taken twice over, is then a degenerate form of the quadric cone in general obtained by the envelope of the polar planes of the points of a line. Prove also that, if the polar planes of all points of a line, with respect to the cubic surface, pass through another line, then two of the four intersections of the first line with the Hessian surface are corresponding points of this surface.

Ex. 3. Let P, P' be any pair of corresponding points of the Hessian, and let the line joining these meet the Hessian again in the points M and N. Let M', N' be the points of the Hessian respectively corresponding to M and N. Prove, (1), that the line $M'N'$ is the intersection of the tangent planes of the Hessian at P and P'; (2), that the polar plane of every point of the line PP', with respect to the cubic surface, contains the line $M'N'$, and (3), that the line $M'N'$ touches the Hessian at M' and N'.

Ex. 4. The cubic surface which is the envelope of the polar planes, in regard to the cubic surface, of the points of a plane, touches the Hessian at the points of a sextic curve. This curve is the locus of the points corresponding to the points of the plane curve in which the given plane cuts the Hessian.

Ex. 5. Let an arbitrary line meet the Hessian in points P, Q, R, S; and let P', Q', R', S' be the points respectively corresponding to these. The polar quadrics of P', Q', R', S', with respect to the cubic surface, are cones, with vertices respectively at P, Q, R, S. Prove that there are two planes, through the line $PQRS$, which are conjugate to one another in regard to each of these four cones. Then, that these two planes, taken with the four planes each containing three of the points P', Q', R', S', are six planes, say $\chi_r = 0$, in terms of which the equation of the cubic surface is expressible in the form $\Sigma c_r \chi_r^3 = 0, (r = 1, 2, \ldots, 6)$. (Reye, *Crelle's J.*, LXXVIII, 1874.)

Ex. 6. Every line of the cubic surface touches the Hessian in two points; these are corresponding points; and the twenty-four points thus arising from the lines of a double-six are the intersections, of the Schur quadric associated with the double-six, with the curve, of order twelve, which is the intersection of the cubic surface and its Hessian. When the cubic surface is $\Sigma a x^3 = 0$, with $\Sigma a x = 0$, in Sylvester's form, the fifty-four points so arising satisfy the equation $\Sigma a x^{-3} = 0$.

Ex. 7. If $u = ax + by + cz + dt$, $v = lx + my + nz + pt$, the Hessian of the cubic surface whose equation is $x^3 + y^3 + z^3 + t^3 + u^3 + v^3 = 0$, is given by

$$xyzt + uxyzt\,\Sigma\, a^2 x^{-1} + vxyzt\,\Sigma\, l^2 x^{-1}$$
$$+ uv\,[\Sigma\,(dl - ap)^2 yz + \Sigma\,(bn - cm)^2 xt] = 0.$$

When $d = 0$, so that the four planes $x = 0$, $y = 0$, $z = 0$, $u = 0$ meet in a point, if we put X, Y, Z, U, V, respectively, for p^2x, p^2y, p^2z, p^2u, p^2v, this reduces to the form

$$tV\,[\Sigma\, l^2 YZ + U\Sigma X\,(bn - cm)^2] + (t + V)\,[XYZ + U\Sigma a^2 YZ] = 0.$$

The six common generators of the two cones obtained by equating to zero the coefficients of tV and $t+V$, in this equation, which have a common vertex, are then six intersecting lines lying on the Hessian surface. We infer that the cubic surface in question is not general enough to be given by the Sylvester form, as a sum of five distinct cubes. The line $V=0$, $t=0$ also lies on this Hessian.

Reduced forms for twenty-three cases of cubic surfaces are given by Cayley, *Papers*, VI, p. 359. It may be a curious problem for the reader to enquire what is the fewest number of cubes by which these forms can severally be represented. That any cubic polynomial, in x, y, z, t, may be represented as a sum of cubes only, each of a linear function of x, y, z, t, with proper coefficients, is clear; as a very particular case it may be shewn there is always an unique representation by a selection from the following twenty cubes: four cubes such as x^3; four cubes such as $(y+z+t)^3$; six cubes such as $(y+z)^3$ and six cubes such as $(y-z)^3$.

The representation of a general cubic surface upon a quadric surface. As both a cubic surface and a quadric surface may be represented upon a plane, it follows that a cubic surface can be represented on a quadric surface. Let A, B, C, D, E be five points of a quadric; these lie in threes in ten planes, giving, in all, ten conics lying on the quadric; there are two generators of the quadric at each of the five points; there are also two cubic curves lying on the quadric, containing the five points (above, p. 129). The representation of the cubic surface upon the quadric can be made in such a way that the twenty-seven lines of the cubic surface correspond to the twenty-seven elements of the quadric which we have enumerated, namely 5 points, 10 conics, 10 generators and 2 cubic curves.

For let the cubic surface be given by $\xi\eta\zeta=\xi'\eta'\zeta'$, where ξ, η, ζ are such that $\xi+\eta+\zeta=\xi'+\eta'+\zeta'$ and $a\xi+b\eta+c\zeta=a'\xi'+b'\eta'+c'\zeta'$. Then, for any σ, we have the relation

$$(a+\sigma)\xi+(b+\sigma)\eta+(c+\sigma)\zeta=(a'+\sigma)\xi'+(b'+\sigma)\eta'+(c'+\sigma)\zeta'.$$

There are two values of σ for which

$$(a+\sigma)(b+\sigma)(c+\sigma)=(a'+\sigma)(b'+\sigma)(c'+\sigma);$$

let these be σ_1 and σ_2; put $\xi_1=(a+\sigma_1)\xi$, and $\xi_2=(a+\sigma_2)\xi$, etc. We then have $\xi_r\eta_r\zeta_r=\xi_r'\eta_r'\zeta_r'$, $\xi_r+\eta_r+\zeta_r=\xi_r'+\eta_r'+\zeta_r'$, for $r=1$ and $r=2$. Now take x, y, z, t by the definitions

$$x/t=(\eta'-\zeta)/(\xi-\xi'),\quad y/t=(\zeta'-\xi)/(\eta-\eta'),\quad z/t=(\xi'-\eta)/(\zeta-\zeta'),$$

and let $x'=t-x$, $y'=t-y$, $z'=t-z$. We easily verify that

$$\xi:\eta:\zeta:\xi':\eta':\zeta'=yz':zx':xy':y'z:z'x:x'y,$$

which involve the two relations $\xi\eta\zeta=\xi'\eta'\zeta'$, $\xi+\eta+\zeta=\xi'+\eta'+\zeta'$. The remaining relation which limits $\xi, \eta, \ldots$ requires, however, that

$$ayz'+bzx'+cxy'=a'y'z+b'z'x+c'x'y,$$

and expresses that the point (x, y, z, t) lies on a quadric surface containing the five points $(1, 0, 0, 0)$, $(0, 1, 0, 0)$, $(0, 0, 1, 0)$, $(0, 0, 0, 1)$, $(1, 1, 1, 1)$.

Denote these points, respectively, by A, B, C, D, E, the generators of one system of the quadric which contains these points, respectively, by a_1, b_1, c_1, d_1, e_1, and those of the other system, through these points, respectively, by a_2, b_2, c_2, d_2, e_2. Further, let the cubic curve lying on the quadric, passing through the five points, which meets the generators of the first system each in *one* point be denoted by γ_1, and the cubic curve on the surface, through the five points, which meets the generators of the second system each in one point, be denoted by γ_2. Then, denoting the line of the cubic surface for which $\xi = 0$, $\xi' = 0$ by (ξ, ξ'), etc., the line which lies on the planes $\xi = \xi'$, $\eta = \eta'$, $\zeta = \zeta'$ by (ξ', η', ζ'), that which lies on the planes $\xi_1 = \xi_1'$, $\eta_1 = \eta_1'$, $\zeta_1 = \zeta_1'$ by $(\xi', \eta', \zeta')_1$, and so on, it may be verified that the twenty-seven lines of the cubic surface correspond to the twenty-seven elements of the quadric surface which have been specified, in accordance with the following scheme:

(ξ, ξ'), the point A ; (ξ', η), the plane ABD ;
(ζ, ξ'), the plane CAE ; (ξ, η'), the plane ABE ;
(η, η'), the point B ; (η', ζ), the plane BCD ;
(ζ', ξ), the plane CAD ; (η, ζ'), the plane BCE ;
(ζ, ζ'), the point C.

(ξ', η', ζ'), ABC ; (η', ζ', ξ'), E ; (ζ', ξ', η'), D ; (ξ', ζ', η'), ADE ;
(ζ', η', ξ'), BDE ; (η', ξ', ζ'), CDE ; $(\xi', \eta', \zeta')_1$, γ_2 ; $(\eta', \zeta', \xi')_1$, d_2 ;
$(\zeta', \xi', \eta')_1$, e_2 ; $(\xi', \zeta', \eta')_1$, a_1 ; $(\zeta', \eta', \xi')_1$, b_1 ; $(\eta', \xi', \zeta')_1$, c_1 ;
$(\xi', \eta', \zeta')_2$, γ_1 ; $(\eta', \zeta', \xi')_2$, d_1 ; $(\zeta', \xi', \eta')_2$, e_1 ; $(\xi', \zeta', \eta')_2$, a_2 ;
$(\zeta', \eta', \xi')_2$, b_2 ; $(\eta', \xi', \zeta')_2$, c_2.

A particular double-six of lines of the cubic surface corresponds then to the two rows of elements on the quadric surface given by

$$\begin{matrix} A, & B, & C, & D, & e_2, & e_1 \\ BCD, & CAD, & ABD, & ABC, & \gamma_2, & \gamma_1 ; \end{matrix}$$

and, if the two rows of lines of this double-six be denoted, respectively, by $l_1, \ldots, l_6$ and $m_1, \ldots, m_6$, and the line which meets the two l_r, m_s, and the two l_s, m_r, be denoted by n_{rs}, the respective elements $n_{1h}\,(h = 2 \ldots 6)$, $n_{2k}\,(k = 3 \ldots 6)$, $n_{3r}\,(r = 4 \ldots 6)$, $n_{4s}\,(s = 5, 6)$, n_{56}, are given by

ABE, CAE, ADE, a_1, a_2; BCE, BDE, b_1, b_2; CDE, c_1, c_2; d_1, d_2; E.

Further, the cubic curve of the cubic surface, obtained by the intersection of this with the general quadric containing the lines m_2, m_3, n_{23} (see above, p. 176), is found to correspond to the conic on the quadric surface given by $Px' + Qy' + Rz' = 0$, that is, to the general conic of the quadric surface passing through the point $(1, 1, 1, 1)$. While, the cubic curve of the cubic surface which is the intersection of this with the general quadric containing the lines l_2, l_3, n_{23} corresponds to the sextic curve on the quadric given by its intersection with the cubic surface $Lx'yz + My'zx + Nz'xy = 0$, or, what is the same thing, with the cubic surface given by $Lx^{-1} + My^{-1} + Nz^{-1} + Kt^{-1} = 0$, wherein $L + M + N + K = 0$. As is seen in the following note, this is the general cubic surface with four double points, at the points A, B, C, D, which contains the point $(1, 1, 1, 1)$.

Ex. Determine the representation on the quadric of the other cubic curves of the cubic surface.

Note on the cubic surface with four double points. The present chapter has been devoted to the fundamental properties of the general cubic surface. Many special cubic surfaces have interesting properties; but there seems special reason for devoting some remarks to that surface which has, in fact, the greatest number of possible double points—as it presents itself in various ways. The surface was first studied by Cayley, *Papers*, I, p. 183 (1844).

The equation $x^{-1} + y^{-1} + z^{-1} + t^{-1} = 0$ clearly represents a cubic surface having each of the four points such as $(1, 0, 0, 0)$ as a double point. The six lines joining the pairs of these points lie entirely on the surface; and there are three other lines, each meeting an opposite pair of these six lines, which also lie on the surface. For instance, one of these three lines is given by $y + z = 0$, $x + t = 0$; the plane $x + t = 0$ meets the surface in this line, and in the line $x = 0$, $t = 0$ counting doubly. The three new lines lie in the plane $x + y + z + t = 0$; and are the joins of the three pairs of opposite intersections of the four lines in which this plane meets the four fundamental planes, $x = 0$, etc. The surface contains no other lines than the nine which have been mentioned; if each of the six lines joining two of the double points be counted four times, these nine lines count as twenty-seven.

If we take three quantities, ξ, η, ζ, which we may regard as coordinates in a plane, defined by

$$\xi : \eta : \zeta = x(y + z) : y(z + x) : z(x + y),$$

where x, y, z, t are the coordinates of a point of the cubic surface, so that, as we may easily verify

$$\xi^{-1} : \eta^{-1} : \zeta^{-1} = tx - yz : ty - zx : tz - xy,$$

then we deduce

$$x^{-1} : y^{-1} : z^{-1} : t^{-1} = \xi - \eta - \zeta : \eta - \zeta - \xi : \zeta - \xi - \eta : \xi + \eta + \zeta,$$

which evidently satisfy the equation of the cubic surface. By these formulae, there is established a birational correspondence between the points of the cubic surface and the points of the plane (ξ, η, ζ). In this plane there are four lines $\xi \pm \eta \pm \zeta = 0$, each of which corresponds to one of the double points of the cubic surface, while the lines $\xi = 0$, $\eta = 0$, $\zeta = 0$ may be regarded as corresponding to the three lines of the surface lying in the plane $x + y + z + t = 0$. In general, to the plane cubic curve in which the surface is met by the plane $Ax + By + Cz + Dt = 0$, there corresponds, in the plane, the cubic curve $AYZT + \ldots + DXYZ = 0$, where X, Y, Z, T are, for brevity, written for $\xi - \eta - \zeta, \ldots, \xi + \eta + \zeta$. This latter is the general cubic curve passing through the six intersections of the four lines $X = 0, Y = 0, \ldots$. Conversely, to a general line of the plane, $u\xi + v\eta + w\zeta = 0$, there corresponds, on the cubic surface, the curve in which this is met by the quadric cone

$$(v + w)\,yz + (w + u)\,zx + (u + v)\,xy = 0\,;$$

this curve lies on the cubic surface

$$ux^{-1} + vy^{-1} + wz^{-1} + (u + v + w)\,t^{-1} = 0,$$

which is another cubic surface having the same four double points as the original. The curve of intersection of the original cubic surface with that represented by $ax^{-1} + by^{-1} + cz^{-1} + dt^{-1} = 0$ corresponds to the line, in the plane, given by

$$(d + a - b - c)\,\xi + (d + b - c - a)\,\eta + (d + c - a - b)\,\zeta = 0,$$

that is to say, the portion of this curve of intersection other than the six joins of the four fundamental points, so corresponds. It may be convenient to describe as the fundamental cubic curves, of the original cubic surface, these cubic curves, which are thus given by its intersection with other cubic surfaces having the same double points. Two of these curves have one common point, other than the four double points, through which they all pass; and one of these curves can be drawn through two arbitrary points of the original surface.

There is, upon any one of these fundamental cubic curves, a particular point; in fact the tangent planes of the cubic surface at all points of this curve meet the curve again in the same point, which is the point in question. For the curve given by the intersection of the cubic surface with the surface

$$ux^{-1} + vy^{-1} + wz^{-1} + (u + v + w)\,t^{-1} = 0,$$

this particular point is of coordinates $(x_0, y_0, z_0, 1)$, where

$$x_0 = (w - u)(u - v)/(w + u)(u + v),$$

etc.; and, with this, the statement may be verified directly, the equation of the tangent plane of the surface at (X, Y, Z, T) being $xX^{-2} + \ldots + tT^{-2} = 0$. But the fact will appear from the argument below. Assuming the truth of this, it follows that the quadric cone of vertex $(x_0, y_0, z_0, 1)$, which projects the cubic curve from this point, touches the cubic surface at every point of this curve. Moreover, as x_0, y_0, z_0, are clearly unchanged by the substitution of u^{-1}, v^{-1}, w^{-1}, respectively, for u, v, w, there is another quadric cone of vertex $(x_0, y_0, z_0, 1)$ which touches the cubic surface along a cubic curve, this being the curve lying on the surface

$$u^{-1}x^{-1} + \ldots + (u^{-1} + v^{-1} + w^{-1})\, t^{-1} = 0.$$

These two cubic curves, together, constitute the complete locus of points of the cubic surface whereat the tangent planes pass through $(x_0, y_0, z_0, 1)$, and may be spoken of as *conjugate* fundamental curves. The point $(x_0, y_0, z_0, 1)$ may be taken to be any point of the cubic surface, the ratios of u, v, w being then determinable from the equations

$$u\,(1 + x_0^{-1}) + v\,(1 + y_0^{-1}) + w\,(1 + z_0^{-1}) = 0,$$
$$u^{-1}(1 + x_0^{-1}) + v^{-1}(1 + y_0^{-1}) + w^{-1}(1 + z_0^{-1}) = 0.$$

The cubic surface is, therefore, such that the tangent cone, of order four, drawn to it from any general point of itself, breaks up into two quadric cones, each meeting the surface in a cubic curve.

The cubic surface contains also another important system of curves, those, namely, which correspond to the conics, in the plane (ξ, η, ζ), which touch the four lines $\xi \pm \eta \pm \zeta = 0$. The equation of such a conic is $a^{-1}\xi^2 + b^{-1}\eta^2 + c^{-1}\zeta^2 = 0$, where a, b, c are any constants for which $a + b + c = 0$. The corresponding curve on the cubic surface satisfies the equation

$$a^{-1}x^2 (y + z)^2 + b^{-1}y^2 (z + x)^2 + c^{-1}z^2 (x + y)^2 = 0\,;$$

or, as $tx\,(y + z) = -yz\,(x + t)$, the curve satisfies the equation

$$a^{-1}(y + z)(x + t) + b^{-1}(z + x)(y + t) + c^{-1}(x + y)(z + t) = 0,$$

equivalent with $a^2 (yz + xt) + b^2 (zx + yt) + c^2 (xy + zt) = 0$. This equation represents a quadric surface, passing through the four double points of the cubic surface, and touching the four lines in which the plane $x + y + z + t = 0$ meets the planes $x = 0, \ldots, t = 0$. The curves in which these quadric surfaces meet the cubic surface are then sextic curves, with four double points (at the fundamental points). Two of these curves pass through an arbitrary point of the cubic surface. These curves have the property, as will appear below, or may be verified directly, that the tangent line of the curve, at any point of the curve, is one of the inflexional lines of the surface at this point (see above, p. 197); it meets the surface at this point in three coincident points, and does not meet the surface at any other point. They are, therefore, called inflexional curves; and there are no other inflexional curves on the surface than these.

Consider, now, the relation of three points of the cubic surface which lie on a line. The third of these is such that all plane sections of the surface which contain the first two equally contain the third. A plane section of the cubic surface, however, corresponds, in the plane (ξ, η, ζ), to a cubic curve through the six points of intersection of the lines $\xi \pm \eta \pm \zeta = 0$. Thus, three points of the cubic surface which are in line correspond to three points of the plane, say P, Q, R, which have the property that all cubic curves through the six fundamental points, and through P, Q, necessarily pass through R; the existence of such a property for plane cubic curves has already been remarked (pp. 156, 178). Of such cubic curves, however, a degenerate one consists of the line $\xi + \eta + \zeta = 0$ taken with the conic defined by P, Q and the three intersections of the other three fundamental lines; and there are three other such degenerate cubic curves. The point R is thus the point through which pass the four conics each defined as passing through the three intersections of three of the fundamental lines, and also through P and Q. With a phraseology previously employed, in which P and Q are taken as Absolute points, and the conics are described as circles, R is the focus of the parabola touching the four fundamental lines (Vol. II, p. 82). That is, if the unique conic, of the form $a^{-1}\xi^2 + b^{-1}\eta^2 + c^{-1}\zeta^2 = 0$, touching the four fundamental lines, which touches PQ, be taken, R is the intersection of the two other tangents drawn to this conic, one from P and one from Q. A particular case, which we employ here, is that if P and Q coincide on the line PQ, the cubic curves considered touching this line at the point of coincidence, then R is the point of contact, with the conic touched by the line PQ, of the other tangent drawn to this conic from the point of coincidence. For distinctness, let the conic, of the form $a^{-1}\xi^2 + b^{-1}\eta^2 + c^{-1}\zeta^2 = 0$, (with $a + b + c = 0$), which touches a given line PQ, be called the conic *corresponding* to this line. There is then, upon the line, a particular point, the point of contact of the line with its corresponding conic; and, when P and Q coincide at this point, the point R also coincides with these. Thus we see that, upon any one of the fundamental cubic curves of the cubic surface, there is a particular point, at which the tangent line of the curve meets the surface in three coincident points; this point lies on a particular one of those sextic curves of the surface, previously spoken of, which correspond to the conics $a^{-1}\xi^2 + \text{etc.} = 0$ $(a + b + c = 0)$, and touches this curve at this point. Moreover, through any point of such a sextic curve, there passes a fundamental cubic curve of the cubic surface, touching the sextic curve at this point. That these sextics are inflexional curves, in the sense explained, is clear. We see, also, that, if we consider the third point of intersection, with the cubic surface, of the tangent line of a fundamental cubic curve, at any point of this, the locus of this third point is the inflexional curve of the cubic surface which touches this cubic curve. Denoting the point of a cubic curve of the surface where it touches an inflexional curve as the inflexional point of the curve, and recalling that, through any point of the plane, two conics of the set

$$a^{-1}\xi^2 + \text{etc.} = 0 \quad (a+b+c=0)$$

can be drawn, we see that two of the fundamental cubic curves of the cubic surface can be drawn through any point of the surface to have this point as their inflexional point. In the plane, the two tangents of the conics in question which pass through any point, are conjugate lines in regard to all these conics; and, if their equations be $u\xi + v\eta + w\zeta = 0$, and $u_1\xi + v_1\eta + w_1\zeta = 0$, we have $uu_1 = vv_1 = ww_1$. This relation then holds for the cubic curves in question through any point of the surface; and these curves may be spoken of as conjugate.

To any point, P, of the plane (ξ, η, ζ), there corresponds a cubic curve, passing through the six intersections of the four fundamental lines, which has a double point at P. For, if we determine the locus of the points of contact of the tangents from P to all the conics $a^{-1}\xi^2 + \text{etc.} = 0$ $(a+b+c=0)$, we find, by elimination of u, v, w from the equations

$$u\xi + v\eta + w\zeta = 0, \quad u\xi_0 + v\eta_0 + w\zeta_0 = 0, \quad u^{-1}\xi + v^{-1}\eta + w^{-1}\zeta = 0,$$

wherein (ξ_0, η_0, ζ_0) are the coordinates of P, the equation

$$\xi(\eta\zeta_0 - \eta_0\zeta)^{-1} + \eta(\zeta\xi_0 - \zeta_0\xi)^{-1} + \zeta(\xi\eta_0 - \xi_0\eta)^{-1} = 0,$$

which represents the curve in question, as is easily verified. If P moves on a line, the various cubic curves so found all pass then through the point where the line touches the conic which corresponds to it. Now, such a cubic curve corresponds to a plane section of the cubic surface having a double point, that is, to the section of the cubic surface by a tangent plane. We can thus infer that the tangent planes of the cubic surface, at all the points of one of its fundamental cubic curves, pass through the inflexional point of the curve, as was stated above.

Ex. Another result may be referred to. To any point, (ξ_0, η_0, ζ_0), of the plane, there is associated a conic which is the envelope of the polar lines of the point in regard to all the conics

$$a^{-1}\xi^2 + \text{etc.} = 0 \quad (a+b+c=0);$$

in particular, this conic touches the two lines through (ξ_0, η_0, ζ_0) which are conjugate to one another in regard to all these conics, these being the tangents at (ξ_0, η_0, ζ_0) of the two of these conics which pass through this point. The equation of this conic is

$$(\xi_0\xi)^{\frac{1}{2}} + (\eta_0\eta)^{\frac{1}{2}} + (\zeta_0\zeta)^{\frac{1}{2}} = 0;$$

it may be referred to, here, as *the envelope conic* for this point. Correspondingly, on the cubic surface, there is associated to any point, (x_0, y_0, z_0, t_0), a locus given by the equation

$$[\xi_0 x (y+z)]^{\frac{1}{2}} + \text{etc.} = 0.$$

The rational form of this, if we use the relations

$$tx^2 (y+z) = -xyz (x+t) \text{ and } tyz (x+y)(x+z) = xyz (tx - yz),$$

which follow from the equation of the cubic surface, is at once found to be

$$yz (\eta_0 - \zeta_0)^2 + xt (\eta_0 + \zeta_0)^2 + \text{etc.} = 0.$$

This is the same as $PT = U^2$, where

$$P = x (\eta_0 + \zeta_0)^2 + y (\zeta_0 + \xi_0)^2 + z (\xi_0 + \eta_0)^2, \quad T = x + y + z + t,$$
$$U = x (\eta_0 + \zeta_0) + y (\zeta_0 + \xi_0) + z (\xi_0 + \eta_0);$$

the equation, therefore, represents a quadric cone, passing through the double points of the cubic surface, whose vertex is on the plane $T = 0$, which, moreover, touches this plane. The vertex of the cone is given by

$$xt^{-1} = \frac{(\zeta_0 + \xi_0)(\xi_0 + \eta_0)}{(\zeta_0 - \xi_0)(\xi_0 - \eta_0)}, \; yt^{-1} = \frac{(\xi_0 + \eta_0)(\eta_0 + \zeta_0)}{(\xi_0 - \eta_0)(\eta_0 - \zeta_0)}, \; zt^{-1} = \frac{(\eta_0 + \zeta_0)(\zeta_0 + \xi_0)}{(\eta_0 - \zeta_0)(\zeta_0 - \xi_0)}.$$

Two cones can be drawn to touch the plane $T = 0$, passing through the four double points, with vertex at this point; the other is evidently obtainable by writing $\xi_0^{-1}, \eta_0^{-1}, \zeta_0^{-1}$ for ξ_0, η_0, ζ_0, respectively. The cone meets the cubic surface in a sextic curve with a double point at each of the four double points of the surface; indeed, such a sextic can similarly be shewn to arise from any conic of the plane. The particular sextic in question touches the three lines of the cubic surface which lie in the plane $T = 0$; and, also, touches the two fundamental cubic curves of the cubic surface lying on the enveloping quadric cones drawn to the surface from (x_0, y_0, z_0, t_0). Let this sextic be described, here, as the envelope sextic associated with this point, and these two cubic curves as the associated cubic curves.

From any point, (x_1, y_1, z_1, t_1), of the cubic surface, there can be drawn four lines, each to meet both these cubic curves, other than the four lines to the double points of the surface, and the one line to the point (x_0, y_0, z_0, t_0); for, these two cubic curves are projected from (x_1, y_1, z_1, t_1) by cubic cones, having, in all, nine common generators. It can be shewn, now, that, when (x_1, y_1, z_1, t_1) is on the sextic associated with (x_0, y_0, z_0, t_0), these four lines become two pairs of coincident lines, each of which then meets these cubic curves, respectively, in two points with the property that the tangent lines of these curves at these points lie in a plane. This plane is then a plane of the developable formed by the common tangent planes of the two associated cubic curves; it meets the cubic surface in a plane cubic curve which touches both these cubic curves; this plane cubic section also passes through (x_1, y_1, z_1, t_1), and touches the associated sextic at (x_1, y_1, z_1, t_1). If the tangent line be drawn at any point, *H*, of one of the two associated cubic curves, there are, we know, four tangent lines of the other cubic curve which meet the former tangent line; the plane determined by the former tangent line and one of these four other tangent lines of the associated cubic curve, meets the cubic surface in a plane cubic curve which, as the point *H* of the first cubic varies, envelops the associated sextic.

To prove this result, consider, in the plane (ξ, η, ζ), the points, say (ξ_0, η_0, ζ_0) and (ξ_1, η_1, ζ_1), corresponding, respectively, to the points (x_0, y_0, z_0, t_0), (x_1, y_1, z_1, t_1) of the cubic surface, of which the latter is, at first, taken quite arbitrarily. There are two conics of the system $a^{-1}\xi^2 + \text{etc.} = 0$ $(a+b+c=0)$ in regard to both of which the points (ξ_0, η_0, ζ_0), (ξ_1, η_1, ζ_1) are conjugate to one another. For, if we consider the conic associated with the point (ξ_1, η_1, ζ_1), whose equation is $(\xi_1\xi)^{\frac{1}{2}} + (\eta_1\eta)^{\frac{1}{2}} + (\zeta_1\zeta)^{\frac{1}{2}} = 0$, there are two tangents of this conic passing through the point (ξ_0, η_0, ζ_0); and, $u\xi + v\eta + w\zeta = 0$ being one of these, there is a particular conic of the system $a^{-1}\xi^2 + \text{etc.} = 0$ $(a+b+c=0)$ for which this line is the polar of (ξ_1, η_1, ζ_1), namely $\xi_1^{-1}u\xi^2 + \eta_1^{-1}v\eta^2 + \zeta_1^{-1}w\zeta^2 = 0$. Consider one of these two conics, of the system in question, in regard to which (ξ_0, η_0, ζ_0) and (ξ_1, η_1, ζ_1) are conjugate: denote this by Σ_1. The polar of (ξ_0, η_0, ζ_0) in regard to Σ_1, taken with the two lines through (ξ_0, η_0, ζ_0), say l and l', which are conjugate to one another in regard to all the conics $a^{-1}\xi^2 + \text{etc.} = 0$ $(a+b+c=0)$, determine a self-polar triad in regard to Σ_1; if then the two tangents be drawn to Σ_1 from (ξ_1, η_1, ζ_1), meeting the line l, say, in *P* and *M*, and meeting the line l' in *Q* and *N*, it is easily seen that the two lines *PQ* and *MN* are tangents of Σ_1, which, by the construction, touches *PN* and *QM* (see Vol. II, Ex. 31, p. 58, and Ex. 5, p. 28). The three points, (ξ_1, η_1, ζ_1), *P*, *Q*, thus correspond, on the cubic surface, to a line drawn from (x_1, y_1, z_1, t_1) to meet the two conjugate cubic curves of the cubic surface associated with (x_0, y_0, z_0, t_0), as, likewise, do the three points, (ξ_1, η_1, ζ_1), *M*, *N*. The two other such transversals drawn from (x_1, y_1, z_1, t_1) similarly arise from the other conic, Σ_2, in regard to which (ξ_0, η_0, ζ_0) and (ξ_1, η_1, ζ_1) are conjugate. Now, it may happen that the conics Σ_1, Σ_2 coincide; namely, when the envelope conic associated with (ξ_1, η_1, ζ_1) contains (ξ_0, η_0, ζ_0), that is, when $(\xi_0\xi_1)^{\frac{1}{2}} + (\eta_0\eta_1)^{\frac{1}{2}} + (\zeta_0\zeta_1)^{\frac{1}{2}} = 0$. Then the conics Σ_1, Σ_2 become the same conic, $\alpha^{-1}\xi^2 + \beta^{-1}\eta^2 + \gamma^{-1}\zeta^2 = 0$, with $\alpha = (\xi_0\xi_1)^{\frac{1}{2}}$, etc., as is easily seen. When this is so, the point (ξ_1, η_1, ζ_1)

is on the envelope conic arising from (ξ_0, η_0, ζ_0), and the point (x_1, y_1, z_1, t_1) is on the sextic curve associated with (x_0, y_0, z_0, t_0). This establishes the result which has been stated.

The theorem that the tangent line of a cubic curve on the cubic surface meets the surface again on the inflexional curve touching the cubic curve, proved above, was given by Laguerre, *Nouv. Ann.*, XI, 1872, p. 342 (*Oeuvres*, II, pp. 275, 281). The representation of the inflexional curves by conics touching four lines was remarked by Lie (see *Geom. d. Berührungstransf.*, I, 1896, p. 342). The theorem given in the example above is derived from F. Morley (*An extension of Feuerbach's theorem, Nat. Ac. of Sc., U.S.A.*, 1916 (2), p. 171). For further references in regard to the cubic surface with four double points, the reader may consult the author's *Multiply-periodic functions* (Cambridge, 1907), pp. 139—151, where the surface arises as touching the Kummer surface at every point of its intersection with it; also Neuberg, *Arch. d. Math. u. Phys.*, XI, 1907, p. 228, and Servais, *Bull. Sc. Acad. r. de Belgique*, VIII, 1922, pp. 50—66 and 103—123.

The Steiner quartic surface. We may consider the surface which is dual to the cubic surface just discussed. As the equation of the tangent plane of the cubic surface is $x_0^{-2}x + y_0^{-2}y + \ldots = 0$, the tangential equation of this cubic surface is $l^{\frac{1}{3}} + m^{\frac{1}{3}} + n^{\frac{1}{3}} + p^{\frac{1}{3}} = 0$, and the equation of the quartic surface, dual to the cubic surface, is $x^{\frac{1}{2}} + y^{\frac{1}{2}} + z^{\frac{1}{2}} + t^{\frac{1}{2}} = 0$. This meets each of the planes $x = 0$, $y = 0$, etc., in a conic, at every point of which it touches the plane, any two of these four conics having a point of contact, on the line of intersection of their planes. The surface has also three double lines, such as that given by $y - z = 0$, $t - x = 0$; these meet in the point $(1, 1, 1, 1)$, which is a triple point of the surface. Every tangent plane of the surface meets it in a curve breaking up into two conics; the four intersections of these consist of the point of contact of the plane, and the three points where the plane meets the double lines of the surface. The surface can be represented, on a plane (ξ, η, ζ), by means of the formulae

$$x = (\xi - \eta - \zeta)^2,\ \ y = (\eta - \zeta - \xi)^2,\ \ z = (\zeta - \xi - \eta)^2,\ \ t = (\xi + \eta + \zeta)^2,$$

which are equivalent to

$$\xi\,(t + x - y - z) = \eta\,(t + y - z - x) = \zeta\,(t + z - x - y).$$

Conversely, let $U = 0$, $V = 0$, $W = 0$, $P = 0$, be any four general conics in a plane, expressed in any variables. Then the surface expressed by $X : Y : Z : T = U : V : W : P$ is essentially the same as the above; namely, we can find four linearly independent linear functions of U, V, W, P which, in terms of suitable coordinates in the plane, have the forms $(\xi - \eta - \zeta)^2, \ldots, (\xi + \eta + \zeta)^2$. For, there are two tangential conics which are inpolar to all of $U = 0, \ldots, P = 0$; let coordinates be taken so that the four common tangents of these two tangential conics are $\xi - \eta - \zeta = 0, \ldots, \xi + \eta + \zeta = 0$; the four original conics are then outpolar to the three point-pairs expressed tangentially by $v^2 - w^2 = 0$, $w^2 - u^2 = 0$, $u^2 - v^2 = 0$, these being the opposite pairs of intersections of the four lines. The general conic, $a\xi^2 + 2f\eta\zeta + \ldots = 0$, outpolar to these point-pairs, is $\xi^2 + \eta^2 + \zeta^2 + 2f\eta\zeta + 2g\zeta\xi + 2h\xi\eta = 0$; as each of $\xi^2 + \eta^2 + \zeta^2$, $\eta\zeta$, $\zeta\xi$, $\xi\eta$ is expressible as a linear function of the four squares in question, the result stated is proved. Incidentally, we remark, that, taking X, Y, Z, T proportional to $\eta\zeta$, $\zeta\xi$, $\xi\eta$, $\xi^2 + \eta^2 + \zeta^2$, the equation of the surface is capable of the form $Y^2Z^2 + Z^2X^2 + X^2Y^2 - XYZT = 0$, which puts in evidence the double lines of the surface.

Thus, if U, V, W, P be any four linearly independent homogeneous quadratic functions of three variables, ξ, η, ζ, the envelope of the plane $Ux + Vy + Wz + Pt = 0$ is a cubic surface with four double points. If U_1, U_2, U_3, respectively, denote $\partial U/\partial\xi$, $\partial U/\partial\eta$, $\partial U/\partial\zeta$, and $U_{11}, U_{12}, \ldots$ denote $\partial^2 U/\partial\xi^2$, $\partial^2 U/\partial\xi\partial\eta, \ldots$, and so on, and u_{rs} denote $U_{rs}x + V_{rs}y + W_{rs}z + P_{rs}t$, the point of contact of the plane corresponding to a particular set of values of ξ, η, ζ is given by the three equations $U_r x + V_r y + W_r z + P_r t = 0$, and the locus of this point of contact, that is, the surface enveloped by the plane, is given by the equation, obtained by eliminating ξ, η, ζ from these three equations,

$$\begin{vmatrix} u_{11}, & u_{12}, & u_{13} \\ u_{21}, & u_{22}, & u_{23} \\ u_{31}, & u_{32}, & u_{33} \end{vmatrix} = 0,$$

where $u_{rs} = u_{sr}$. The quadrics obtained by equating to zero the first minors of this determinant have four points in common; these are the double points of the cubic surface.

Ex. 1. For the general cubic surface, expressed, in Sylvester's way, by $\Sigma ax^3 = 0$, with $\Sigma ax = 0$, the envelope of the polar planes of the points of the plane which passes through the three points $(x_1, y_1, \ldots)$, $(x_2, y_2, \ldots)$, $(x_3, y_3, \ldots)$, (cf. p. 209), is expressible in the form above, with

$$u_{11} = axx_1^2 + \ldots + euu_1^2, \text{ and } u_{12} = axx_1x_2 + \ldots + euu_1u_2, \text{ etc.}$$

Ex. 2. It follows from the dual relation of the cubic surface with four double points, and the Steiner quartic surface, or may be proved directly, that, by the representation upon the plane, the inflexional curves of the Steiner surface are made to correspond to the conics, expressed by $a^{-1}\xi^2 + \text{etc.} = 0$ $(a + b + c = 0)$, which are inpolar to the conics to which the coordinates on the surface are proportional.

Ex. 3. The inflexional curves of the Steiner surface are rational quartic curves, having the three double lines of the surface as chords. Of the four points in which any such curve is met by a tangent plane of the surface, regarded as represented by the values of the parameter by which any point of the curve may be rationally expressed, two are harmonic conjugates in regard to the other two.

Ex. 4. The inflexional curves of the surface expressed by the equation $x^m + y^m + z^m + t^m = 0$ satisfy the equation

$$(ax^m)^{\frac{1}{2}} + (by^m)^{\frac{1}{2}} + (cz^m)^{\frac{1}{2}} = 0,$$

where a, b, c are any constants whose sum is zero. Verify this in the case of the cubic surface with four nodes and in the case of the Steiner surface.

Ex. 5. If $\phi_1 = 0, \ldots, \phi_6 = 0$ be any six conics in a plane, the equations $x_1/\phi_1 = x_2/\phi_2 = \ldots = x_6/\phi_6$ represent, in space of five dimensions, a surface (containing ∞^2 points), known as Veronese's surface. Evidently the Steiner surface may be regarded as arising by projection of this surface, *from a line.*

Ex. 6. The lines joining a point, O, of coordinates x, y, z, t, to the points $(a, 1, 1, 1)$, $(1, b, 1, 1)$, etc., meet the planes $x=0$, $y=0$, etc., respectively, in coplanar points. Prove that the locus of O is the Hessian of the cubic surface

$$\Sigma \frac{x^3}{a(a-1)} + \frac{u^3}{(e-1)(3-2e)} = 0,$$

where e and u are defined by

$$1+\Sigma(a-1)^{-1}+(e-1)^{-1}=0, \quad \Sigma x(a-1)^{-1}+u(e-1)^{-1}=0.$$

When $a=b=c=d=-3$, taking for Absolute Conic the intersection of the plane $x+y+z+t=0$ with the quadric $x^2+y^2+z^2+t^2=0$, shew (after Steiner, 1845) that the cubic surface with four double points is the locus of a point such that the feet of the four perpendiculars drawn from it to the four planes $x=0$, $y=0$, etc. are coplanar. See Jessop, *Quartic Surfaces*, p. 190; also p. 36 above. And *Archiv d. Math. u. Phys.* IV, 1903, p. 275 (a reference for which the writer is indebted to Mr J. P. Gabbatt).

For the Steiner surface, the reader may consult Jessop, *Quartic Surfaces* (Cambridge, 1916), Chapter VII. For Veronese's surface, see Cayley, *Papers*, VI, p. 198 (1868), and Bertini, *Geometria projettiva d. iperspazi*, 1907, Chapters XIV, XV. See also Castelnuovo, *Atti d. r. Acc. d. Lincei*, III, 1894, p. 22, who proves, after Kronecker, that an irreducible algebraic surface which is met by ∞^2 planes in reducible curves, if not a ruled surface, is a Steiner quartic surface.

CORRECTIONS FOR VOLUMES I AND II

VOLUME I.

Page 28, line 7. *For* FCF′ *read* FRF′.
p. 39, l. 22. *For* S_{2n-1} *read* S_{2n-2}.
p. 47, l. 12 f. b. *For* insection *read* intersection.
p. 48, l. 8 f. b. *For* (2) *read* (3).
p. 65, l. 13. *For* b + a *read* 0 + a.
p. 68, ll. 2, 3. Both diagonal elements should be $\alpha\beta - \alpha'\beta'$.
p. 74, l. 18. *For* meet *read* meets.
p. 75, l. 20. *For* P = 0 *read* P = O.
p. 89, l. 16. *For* C − mC′ *read* C − C′.
p. 92, l. 10 f. b. *For* $\sigma\rho\sigma^{-1}B'$ *read* $\sigma\rho^{-1}\sigma^{-1}B'$.
p. 99, l. 11. *For* A *read* A′.
p. 102, l. 11 f. b. *For* any point *read* any point except B or C.
p. 108, l. 2 f. b. *For* B″ *read* B′.
p. 127, l. 14. *For* P and V *read* P_0 and V.
p. 155, l. 5 f. b. After *a*, *b*, *c*, *add* and A, B, C are not in line,
p. 156, l. 3 f. b. When *a*, *b*, *c* meet in a point, and A, B, C are in line, there *may* be an infinite number of solutions.
p. 174, ll. 16, 17 f. b. Interchange O′, U′, W′ and O_1', U_1', W_1'.
p. 174, last line. *For* $m_1' + n_2' - n_2'$ *read* $m_1' + m_2' - n_2'$.

VOLUME II.

p. 13, l. 8. *For* MacClaurin *read* Maclaurin.
p. 28, l. 16. *For* PH, PK *read* PQ, PQ′.
p. 28, l. 18. *For* pole *read* polar.
p. 49, l. 11 f. b. *For* C *read* O.
p. 58, l. 1. *For* intersecting *read* interesting.
p. 65, l. 13 f. b. *For* pole *read* polar.
p. 83, last l. *For* hvperbolas *read* hyperbolas.
p. 86, l. 2 f. b. *For* from (S or *read* from S (or.
p. 92, l. 17 f. b. *For* Σ, in A, B, C, D′ *read* Σ, and meets Σ in A, B, C, D′.
p. 96, l. 10 f. b. *For* any algebraic symbol *read* any algebraic symbol other than zero.
p. 102, l. 19. *For* we can *read* we can if L contain x.
p. 114, l. 18. See Morley, *Bull. Amer. Math. Soc.* I, 1895, p. 116, and *Proc. Camb. Phil. Soc.* XX, 1920, p. 119.
p. 119, last l. *Read*

$$a^2l^2 + b^2m^2 - n^2 + \lambda(l^2 + m^2).$$

p. 120, l. 1. *Read* a^2l^2 *for* a^2b^2.
p. 125, last l. *Read* bc′ + b′c − 2ff′.
p. 130, l. 15. *For* S *read* S_1.
p. 140, l. 7. *For* conic *read* conics.
p. 167, l. 3 f. b. *For* his *read* this.
p. 169, l. 24. *For* $B^2 + C^2 - A^2 - D^2$ *read* $A^2 + D^2 - B^2 - C^2$, and so in ll. 25, 26. In the first terms of y_1 and z_1 *read* x *for* y and z.
p. 197, l. 23. *For* lines *read* segments.
p. 203, l. 13 f. b. *For* $-(A + \phi)$ *read* $-(A + \psi)$.
p. 205, l. 7 f. b. *Read* $\left(1 - \frac{G^2}{CA}\right)^{\frac{1}{2}}$.
p. 212. In the left-hand diagram *join* the points A, E.
p. 231, l. 16. *For* three *read* the.
p. 234, l. 4. *For* (QR, P′Q) and (RP, Q′R) *read* (QR, P′Q′) and (RP, Q′R′), respectively. And so in ll. 5, 6.
p. 234, l. 5 f. b. *For* X′U *read* X′U′.
p. 237. Add reference to Cayley, 'Sixth Memoir on Quantics,' *Papers*, II, p. 583. Also to Beltrami, 'Saggio di interpret. d. Geom. non-eucl.,' *Giorn. d. Mat.* VI, p. 284 (*Opere*, I, p. 374), 1868; and 'Theoria fond. d. spazii di curv. cost.,' *Ann. d. Mat.* II, p. 232 (*Opere*, I, p. 406).
p. 238, l. 14 of ref. to p. 41, in Y (− D + D′ + A′, B′, C′, D′, 0), *for* second D′ *read* D.

The author desires to acknowledge the kindness of Miss H. P. Hudson, London, of Professor Carslaw, Sydney, and especially of Professor Wilton, Adelaide, for the majority of these corrections.

INDEX

CAMBRIDGE: PRINTED BY W. LEWIS AT THE UNIVERSITY PRESS

Printed in the United States
By Bookmasters